Sequence-Based Classification of Select Agents

A BRIGHTER LINE

Committee on Scientific Milestones for the Development of a Gene Sequence-Based Classification System for the Oversight of Select Agents

Board on Life Sciences

Division on Earth and Life Studies

NATIONAL RESEARCH COUNCIL
OF THE NATIONAL ACADEMIES

THE NATIONAL ACADEMIES PRESS
Washington, D.C.
www.nap.edu

THE NATIONAL ACADEMIES PRESS 500 Fifth Street NW Washington, DC 20001

NOTICE: The project that is the subject of this report was approved by the Governing Board of the National Research Council, whose members are drawn from the councils of the National Academy of Sciences, the National Academy of Engineering, and the Institute of Medicine. The members of the committee responsible for the report were chosen for their special competences and with regard for appropriate balance.

This study was supported by Contract No. NO1-OD-4-2139 (Task Order #216) between the National Academy of Sciences and the National Institutes of Health. The content of this publication does not necessarily reflect the views or policies of the National Institutes of Health, nor does mention of trade names, commercial products, or organizations imply endorsement by the U.S. government.

International Standard Book Number-13: 978-0-309-15904-3
International Standard Book Number-10: 0-309-15904-0

Additional copies of this report are available from the National Academies Press, 500 Fifth Street NW, Lockbox 285, Washington, DC 20055; (800) 624-6242 or (202) 334-3313 (in the Washington metropolitan area); Internet, http://www.nap.edu.

Printed in the United States of America.

THE NATIONAL ACADEMIES

Advisers to the Nation on Science, Engineering, and Medicine

The **National Academy of Sciences** is a private, nonprofit, self-perpetuating society of distinguished scholars engaged in scientific and engineering research, dedicated to the furtherance of science and technology and to their use for the general welfare. Upon the authority of the charter granted to it by the Congress in 1863, the Academy has a mandate that requires it to advise the federal government on scientific and technical matters. Dr. Ralph J. Cicerone is president of the National Academy of Sciences.

The **National Academy of Engineering** was established in 1964, under the charter of the National Academy of Sciences, as a parallel organization of outstanding engineers. It is autonomous in its administration and in the selection of its members, sharing with the National Academy of Sciences the responsibility for advising the federal government. The National Academy of Engineering also sponsors engineering programs aimed at meeting national needs, encourages education and research, and recognizes the superior achievements of engineers. Dr. Charles M. Vest is president of the National Academy of Engineering.

The **Institute of Medicine** was established in 1970 by the National Academy of Sciences to secure the services of eminent members of appropriate professions in the examination of policy matters pertaining to the health of the public. The Institute acts under the responsibility given to the National Academy of Sciences by its congressional charter to be an adviser to the federal government and, upon its own initiative, to identify issues of medical care, research, and education. Dr. Harvey V. Fineberg is president of the Institute of Medicine.

The **National Research Council** was organized by the National Academy of Sciences in 1916 to associate the broad community of science and technology with the Academy's purposes of furthering knowledge and advising the federal government. Functioning in accordance with general policies determined by the Academy, the Council has become the principal operating agency of both the National Academy of Sciences and the National Academy of Engineering in providing services to the government, the public, and the scientific and engineering communities. The Council is administered jointly by both Academies and the Institute of Medicine. Dr. Ralph J. Cicerone and Dr. Charles M. Vest are chair and vice chair, respectively, of the National Research Council.

www.national-academies.org

COMMITTEE ON SCIENTIFIC MILESTONES FOR THE DEVELOPMENT OF A GENE SEQUENCE-BASED CLASSIFICATION SYSTEM FOR THE OVERSIGHT OF SELECT AGENTS

JAMES W. LEDUC (*Chair*), Professor, Microbiology and Immunology; Chair, Global Health; and Director, Galveston National Laboratory, The University of Texas Medical Branch at Galveston
RALPH BARIC, Associate Research Professor, Microbiology and Immunology, University of North Carolina at Chapel Hill School of Public Health
ROGER G. BREEZE, Chief Executive Officer, Centaur Science Group
R. MARK BULLER, Professor, Department of Molecular Microbiology and Immunology, Saint Louis University School of Medicine
SEAN R. EDDY, Group Leader, Janelia Farm, Howard Hughes Medical Institute
STANLEY FALKOW, Robert W. and Vivian K. Cahill Professor, Microbiology and Immunology; Geographic Medicine; and Infectious Diseases, Stanford University School of Medicine
RACHEL E. LEVINSON, Director, National Research Initiatives, Office of the Vice President for Research and Economic Affairs, Arizona State University
JOHN MULLIGAN, Chairman and Chief Scientific Officer, Blue Heron Biotechnology
ALISON D. O'BRIEN, Professor and Chair, Department of Microbiology and Immunology, Uniformed Services University of the Health Sciences
FRANCISCO OCHOA-CORONA, Assistant Professor, National Institute for Microbial Forensics & Food and Agricultural Biosecurity (NIMFFAB), Department of Entomology and Plant Pathology, Oklahoma State University
JANE S. RICHARDSON, James B. Duke Professor, Department of Biochemistry, Duke University Medical Center
MARGARET RILEY, Professor of Biology, Department of Biology, University of Massachusetts
TOM SLEZAK, Program Leader, Global Security Program, Lawrence Livermore National Laboratory

Staff

INDIA HOOK-BARNARD, Study Director and Program Officer
CARL-GUSTAV ANDERSON, Senior Program Assistant

Preface

Recent studies by the National Research Council have focused on the tension between the rapid advances in biotechnology that clearly benefit humankind and the potential use of the same advances for nefarious purposes. The 2004 report, *Biotechnology Research in an Age of Terrorism* helped to focus attention on the issue, and among other recommendations called for the creation of a National Science Advisory Board for Biodefense (NSABB) to serve as a bridge between the government and scientific communities in raising awareness of the potential for misuse of biotechnology. A later report, *Globalization, Biosecurity, and the Future of the Life Sciences* carried the discussion forward with a global perspective and promoted a global common culture of awareness and a shared sense of responsibility among life scientists. In 2006, the NSABB issued *Addressing Biosecurity Concerns Related to the Synthesis of Select Agents,* which called for expert evaluation to determine whether an alternative framework based on predicted features and properties encoded by nucleic acids, such as virulence or pathogenicity, can be developed and used in lieu of the current finite list of specific agents and taxonomic definitions. Our committee was tasked with identifying "the scientific advances that would be necessary to permit serious consideration of developing and implementing an oversight system for Select Agents that is based on predicted features and properties encoded by nucleic acids rather than a relatively static list of specific agents and taxonomic definitions."

Our committee was populated with persons who had expertise in several complementary fields. The amalgamation of scientific backgrounds allowed us to address our task from different viewpoints and to assess the potential impact of our recommendations on various sectors. We benefited by having committee members who were experts in human and animal health and leaders in the development of policy relevant to these fields; leaders in fundamental structural, evolutionary, and computational biology and bioinformatics; scientists dedicated to the study of pathogenic viruses and bacteria; and experts from the commercial biotechnology sector. Many committee members were personally involved either directly or indirectly in research on plant or animal pathogens designated as Select Agents and thus had first-hand experience in dealing with the relevant regulations and security requirements implemented in recent years to reduce the risk of misuse. We were especially fortunate to have a senior scientist and executive in the biotechnology industry who was able to offer a unique perspective on the role of industry in implementation of current and future steps that might be taken to reduce the risk of misuse of synthetic biology.

We were informed by the shared expertise of many professionals in synthetic biology, security, public health, human and animal medicine, the life sciences, informatics and several other relevant fields as we grappled with our challenging task. Specifically, Julia Kiehlbauch, Robbin Weyant, Claudia Mickelson, Edward You, and Amy Patterson helped us to understand the current structure for oversight of Select Agents. Peter Pesenti, John Mulligan, Marcus Graf, Claes Gustafsson and Stephen Maurer discussed with us the current mechanisms and criteria for screening and surveillance at the sequence level. Stanley Falkow, Jeffrey Taubenberger, Michael Katze, Ralph Baric and Ramon Felciano discussed virulence; and Sean Eddy, Jonathan Eisen, Elliot Lefkowitz, John Moult and Ian Lipkin addressed gaps, challenges and potential milestones in predicting pathogenicity from sequence information. In addition, Carol Linden, Arturo Casadevall, David Relman, Mary Groesch, Jacqueline Corrigan-Curay and James Blaine all met personally with our committee and joined in our discussions. We sincerely thank all those who took time from their busy schedules to meet with us, answer our questions, and guide us in our deliberations.

Early in our discussions, it became apparent that the criteria historically used to designate a pathogen as a Select Agent included characteristics that cannot be determined by sequence alone and therefore cannot be predicted with the degree of certainty required for regulatory purposes. We soon concluded that a sequence-based *prediction* system for oversight of Select Agents is not now possible, nor is it likely to be feasible in the foreseeable future. We did, however, recognize that a sequence-based *classification* system for Select Agents focused on consideration of sequences of concern could be developed and might help to clarify taxonomic distinctions among recognized Select Agents. By focusing on "sequences of concern" and coupling that with a cautionary alert (a "yellow flag system"), one might effectively address both biosecurity and biosafety goals. Near-term milestone and long-term research objectives were defined and are discussed in our report. Throughout our deliberations, we continually tried to balance the need for safety and security, while recognizing the challenges of potential dual-use applications that arise as the scientific community improves its understanding of the genomic basis that leads one organism to be pathogenic, while its near neighbor is not. We were also concerned about the potential burden that such an oversight program might have on the day-to-day conduct of science and the biotechnology business sector, and about the opportunities that might be missed. We concluded that a gene sequence-based classification system *could* be developed. We did not, however, address whether such as system *should* be developed or whether the additional administrative structure needed to maintain such a system would be justified. Therefore, we do not specifically recommend that either the classification or yellow flag system be implemented. Rather, we provide information about what is technologically feasible, and emphasize that the potential benefits of such a system should be considered and weighed against the cost and complexity of implementation.

The Nobel laureate Joshua Lederberg was quoted in Richard Preston's 1998 article on bioweaponeers in *The New Yorker* as saying: "There is no technical solution to the problem of biological weapons. It needs an ethical, human, and moral solution if it's going to happen at all. Don't ask me what the odds are for an ethical solution, but there is no other solution. But would an ethical solution appeal to a sociopath? (Preston 1998)" We find ourselves today, more than a decade later, still searching for a technical solution to a challenge that has grown beyond biological warfare and now encompasses the threat of bioterrorism as well; a challenge that is ever more complex and threatening as biotechnology advances and access to it expands. We can attempt to harness technology to lessen risks, but we would be wise to heed Lederberg's advice to couple this with efforts toward an "ethical, human and moral solution."

The committee wishes to express its sincere thanks and appreciation to India Hook-Barnard, our study director and program officer, for her leadership, guidance and expertise, coupled with good nature and charming personality. We benefited greatly from her dedication and creativity throughout the study. She was ably assisted by Carl-Gustav Anderson, senior program assistant, who ensured that our every need was met during our meetings and conference calls and worked diligently to coordinate schedules in what must have been a nearly impossible task. Our project was expertly guided by Fran Sharples as the director of the Board on Life Sciences.

Finally, on a personal note, I would like to express my sincere appreciation to the members of the committee, who generously donated their time and knowledge to make this project both extremely productive and very enjoyable. Our discussions were frank, open, and honest, and they benefited greatly from the diversity of our backgrounds and our complementary experiences. We were indeed more than the sum of our own individual contributions. It has been my pleasure and privilege to work with each of you.

James W. Le Duc, Chair

Acknowledgments

This report has been reviewed in draft form by individuals chosen for their diverse perspectives and technical expertise in accordance with procedures approved by the National Research Council's Report Review Committee. The purpose of this independent review is to provide candid and critical comments that will assist the institution in making its published report as sound as possible and to ensure that the report meets institutional standards of objectivity, evidence, and responsiveness to the study charge. The review comments and draft manuscript remain confidential to protect the integrity of the deliberative process.

We thank the following individuals for their review of this report:

Ken Berns, *Florida College of Medicine*
Arturo Casadevall, *Yeshiva University*
Rocco Casagrande, *Gryphon Scientific*
David Eisenberg, *University of California, Los Angeles*
Tanja Kortemme, *University of California, San Francisco*
Gigi Kwik-Gronvall, *University of Pittsburgh*
Henry Metzger, *National Institutes of Health*
Claudia Mickelson, *Massachusetts Institute of Technology*
Stephen Ostroff, *Pennsylvania Department of Health*
Julian Parkhill, *The Sanger Institute*
Harvey Rubin, *University of Pennsylvania*
Janet Shoemaker, *American Society for Microbiology*
Frederick Sparling, *University of North Carolina*

Although the reviewers listed above have provided many constructive comments and suggestions, they were not asked to endorse the conclusions or recommendations, nor did they see the final draft of the report before its release. The review of the report was overseen by **Ronald Atlas**, *University of Louisville* and **Adel Mahmoud**, *Princeton University*. Appointed by the National Research Council, they were responsible for making certain that an independent examination of this report was carried out in accordance with institutional procedures and that all review comments were carefully considered. Responsibility for the final content of the report rests entirely with the authoring committee and the institution.

We thank Dr. Louise Teel and Dr. Christy Ventura of the O'Brien lab; Alisha Prather of Galveston National Lab; and Dr. Jeremy Block of the Richardson lab, for thoughtful suggestions and contributions to the Appendices.

We are grateful to those who attended and participated in the September 3, 2009, Workshop (Appendix D) and those who met with the committee during the course of this study. These individuals, named in the preface, were generous with their time, expertise, and ideas, and were helpful to the committee's work.

Contents

Summary

In 2006, the National Science Advisory Board for Biosecurity (NSABB) released a report, *Addressing Biosecurity Concerns Related to the Synthesis of Select Agents* (NSABB 2006), which considered the effects of synthetic biology and DNA synthesis technology on biosecurity and the current Select Agent Regulations. The principal concerns that it addressed were that

- DNA synthesis technology is rapidly diminishing barriers to acquisition of pathogens, because an increasing variety of organisms may be instantiated by whole genome synthesis, rather than by transfer of samples of existing organism stocks or cultures;
- Natural variation and intentional genetic modification blur the boundaries around any discrete list based on taxonomic names
- Synthetic biology may enable the accidental or deliberate creation of entirely novel pathogens unrelated to current ones.

One of the NSABB recommendations proposed that

> a group of experts from the scientific community be assembled to determine if an alternative framework based on predicted features and properties encoded by nucleic acids, such as virulence or pathogenicity, can be developed and utilized in lieu of the current finite list of specific agents and taxonomic definitions. (NSABB 2006)

Thus, the present study was initiated with the title "Scientific Milestones for the Development of a Gene Sequence-Based Classification System for Oversight of Select Agents" on the basis of this recommendation. The committee was specifically charged with identifying:

> the scientific advances that would be necessary to permit serious consideration of developing and implementing an oversight system for Select Agents that is based on predicted features and properties encoded by nucleic acids rather than a relatively static list of specific agents and taxonomic definitions. (Appendix A)

It is implicit in the charge that a "predictive oversight" system is not now feasible. It is also implicit that "gene sequence-based classification," is synonymous with "predict[ing] features and properties encoded by nucleic acids." However, it soon became clear that the committee was confronted by two quite different tasks, one of which is feasible and one is not. It is possible to *classify* a new sequence as belonging within a group of known sequences; it is *not* feasible to *predict* the function(s) that sequence encodes. **Thus, it is essential to distinguish sequence-based *classification* from sequence-based *prediction* of biological function.**

A sequence-based prediction *system for oversight of Select Agents is not possible now and will not be possible in the usefully near future.*

- Select Agent is not a biological term; rather it is a regulatory designation. Some properties historically considered in assigning an organism to the Select Agent list are not biological properties, and therefore, can never be determined from the organism's genome sequence.
- High-level biological phenotypes—such as pathogenicity, transmissibility, and environmental stability—cannot plausibly be predicted with the degree of certainty required for regulatory purposes, either now or in the foreseeable future.
- Reliable prediction of the hazardous properties of pathogens from their genome sequence alone will require an extraordinarily detailed understanding of host, pathogen, and environment interactions integrated at the systems, organism, population, and ecosystem levels. It is a prediction problem of the greatest complexity.
- Biology is not binary. Microorganisms are not either "potential weapons of mass destruction" or "of no concern." No single characteristic makes a microorganism a pathogen, and no clear-cut boundaries that separate a pathogen from a non-pathogen. Pathogenic microorganisms are not defined by taxonomy; it is common for a given microbial species to have both pathogenic and non-pathogenic representatives. An agent has multiple biological attributes, and the degree to which these are expressed fall along a spectrum for each biological characteristic;[1] consequently, agents present varying degrees of risk.

[1]For example, one microorganism may be highly virulent, but poorly transmissible from person to person, whereas another agent may spread easily, but produce only mild illness.

- For the foreseeable future, the only reliable predictor of the hazard posed by a biological agent will be actual experience with that agent.

Synthetic genomics and the natural complexity of biology increasingly present challenges to biosecurity and biosafety that need to be addressed well before prediction of biological function will be feasible.

- There is a need to provide increased clarity—for investigators, biohobbyists, synthesis companies, and law enforcement officials—about which DNA sequences are subject to the Select Agent Regulations and which are not.
  - Currently, the boundaries around the taxonomic names of Select Agents on the list are unclear. How similar should two sequences be for them to be given the same name?
  - It is also unclear how much (which parts) of an agent must be present for it to be considered a Select Agent. When should a sequence be regarded as a non-functional "genomic fragment" as opposed to a "complete" agent subject to the Select Agent Regulations?
- It might also be desirable to provide information and oversight for "sequences of concern" that are not themselves Select Agents, but potentially could be used to produce a threat.
  - To make it harder for people with nefarious intent to develop pathogens or toxins as weapons or as tools for bioterror without detection.
  - To avoid the accidental, inadvertent, or ill-advised production of hazardous constructs by well-meaning investigators.

A gene sequence-based classification *system for Select Agents and a yellow flag biosafety system for "sequences of concern" could be developed with current technologies.*[2]

- A classification system could provide much needed clarification regarding application of the Select Agent Regulations.
  - For the purposes of regulation, a discrete taxonomic list of Select Agents, augmented by sequence-based classification to better circumscribe taxonomic distinctions blurred by natural and synthetic variation and modification, is a reasonable strategy to maintain for the foreseeable future.
  - Sequence-based classification is strictly operational—a set of tools for drawing decision boundaries around known sequences that do or do not belong to a desired classification. Those tools are used

[2]As noted throughout this report, the classification and "yellow flag" system are presented as proposals for consideration; they should not be read as recommendations.

now for robust and automatic classification of gene sequences into usefully annotated sequence families.

- An operational definition of a complete Select Agent would not predict whether a sequence encodes a functional pathogen. Sequence-based classification strategies would more sharply define the Select Agent Regulations to deal with issues raised by DNA synthesis and natural variation, and would thus establish a "brighter line": an unambiguous procedure for deciding when a genome sequence is assigned one of the taxonomic names on the Select Agent list.
- The problem of classifying a sequence as a complete Select Agent genome (subject to the Select Agent Regulations) has two dimensions: (a) *content*[3]—how much sequence (how many parts) must be present to distinguish a potentially complete "infectious form" of an agent from a non-covered "genomic fragment" or "non-infectious component"; and (b) *distance*—how similar must the sequences of each of those parts be to an actual Select Agent sequence for the same Select Agent taxonomic name to be assigned to the synthetic organism.
- For each Select Agent, given a minimal parts list (*content*) and a profile-based classification system for each part (*distance*), the classification system could be tested, benchmarked, and challenged against known genome sequences. Once developed, the system could be updated to reflect the state of the art of biology and computation and to be correctly harmonized with the Select Agent list.

- A "yellow flag" biosafety system could provide a means of guidance and oversight for "sequences of concern."
  - The yellow flag system would function as an extension of biosafety; however, because it is not regulatory, it could also provide information relevant to biosecurity in a more dynamic and timely fashion than the Select Agent Regulations.
  - The best way to deal with the unquantifiable threat of novel synthetic pathogens is through enhancements to the laboratory and clinical biosafety measures already established for dealing with the threat of emerging natural pathogens.
  - The yellow flag system would comprise four main elements; a centralized biosafety sequence database, annotation of the sequences as empirical evidence of the function of the genes encoded by the sequences is acquired, a process for review and assessment of the

[3]As discussed in Chapter 3, content could be defined by a single gene, such as in the case of a regulated toxin.

evidence to determine the disposition of sequence of concern, and a yellow flag for sequences that are deemed "of concern."

The sequence-based classification presented by the committee is technologically feasible and may improve the current system; however, such a system does have limitations and potential adverse consequences. Therefore, we do not specifically recommend that it be implemented. Rather, we make two recommendations:

- **The sequence space around each discrete taxonomic name on the Select Agent list needs to be clearly defined, so that Select Agent status can be unambiguously determined from a genome sequence (for example, by a DNA synthesis company).**
  The sequence space should be broad enough to include the plausible modifications and chimeras that experts reasonably believe probably also act as Select Agents, without encompassing existing non-Select Agents.
- **A sequence-based classification system could address this problem, and should be considered and weighed against the cost and complexity of implementing this technological augmentation to the current Select Agent Regulations.**

The committee identified specific milestones or focus areas that would aid in developing and implementing a sequence-based classification system and could yield information to improve prediction of function from sequence and to enhance understanding of infectious disease.

- Near-term milestones include:
  - *A sequence database with a Select Agent focus.* A necessary precondition of a classification system is to have a number of representative sequences that belong to each desired classification, and a number of the most closely related sequences that do not belong. A comprehensive sequence database would thoroughly cover naturally occurring genetic variation based on geographic distribution, ecological or laboratory adaptations, and those associated with clinical severity or attenuation. The database would include not only Select Agent sequences, but also a representative set of near-neighbors for each Select Agent.
  - *An expanding sequence database of all biology.* There are massive gaps in our knowledge of the genetic characteristics of much of the biological world. Such a sequence database could be used to help to identify "sequences of concern" that may be appropriate

to monitor in the yellow flag system, in the interests of biosafety or biosecurity.
    - *Stratification of the Select Agent list.* Several recent advisory panels have recommended stratification or reduction of the Select Agent list, to "focus the highest scrutiny on those agents that are indeed of greatest concern" and we are in agreement with that recommendation. Prioritizing the Select Agent list on the basis of risk would make any sequence-based approach to oversight more feasible.
- Long-term areas of research include:
    - Protein structure and function;
    - Gene expression and regulation;
    - Pathogenic mechanisms
    - Animal models of disease
    - Data and information management for systems biology
    - Synthetic biology
    - Metagenomics and phylogenomic, including the human microbiome

The near-term milestones and long-term research aim either to expand the general frontiers of biological knowledge or to apply existing knowledge to the Select Agent Regulations. Our committee was deeply uncomfortable with research programs that would seek to expand knowledge solely for the purpose of improving the Select Agent Regulations.

Developing the ability to predict Select Agent pathogenicity from genome sequence raises serious dual-use concerns, because prediction and design go hand in hand. Accurate computational prediction of Select Agent characteristics from genome sequences enables computational design and optimization of bioweapon genome sequences. Predicting phenotype from genotype and improving public health by increasing our understanding of pathogenicity are two major goals of biology. It does not seem wise to make *special* plans for an effort in predicting the characteristics of Select Agents, in advance of other important frontiers of biological knowledge.

It is more prudent to base the Select Agent Regulations on the current state of biological knowledge, as an applied problem, not a basic research problem. Predictive successes in the general biology research community should be passively monitored. Once biology in general approaches the goal of determining pathogenicity from sequence, then it would be appropriate to consider putting in place a predictive oversight system to identify Select Agent properties from a novel genome sequence. That time may not come for decades, and may be more than a century away.

In the meantime, the technology and knowledge base for sequence-based classification exist now. Even a classification system can present dual-use issues, because implementing the system usefully requires that the information be shared. Listing the "parts" of a Select Agent and identifying other "sequences of concern" entirely on the basis of their potential to be dangerous when incorporated into a synthetic construct disseminates knowledge that could theoretically facilitate the design of a synthetic pathogen by a "bad actor." However, inasmuch as the knowledge would be based on the current published state of the art (and on pathogen sequences that are already widely available in Genbank), any additional dual-use concerns are not nearly as grave.

The Select Agent Regulations strive to balance a need for regulating access to the most dangerous pathogens with the need to minimize the regulatory burden on basic biological research aimed at monitoring, understanding, treating, and preventing disease. If the Select Agent Regulations are too burdensome, they may diminish long-term safety. Our report stops short of recommending the implementation of any specific sequence-based system for defining Select Agents; it was not in our charge, and we were not properly constituted to estimate the costs, benefits, or risks associated with any specific implementation program. We do find that the sequence-based classification system and yellow flag system are technologically feasible, but we have not carefully examined their cost or their effects on basic research or national security. We have made no argument that the positive aspects of using such systems to clarify a sequence-based definition of the discrete taxonomic names on the Select Agent list would outweigh any negative aspects of adding layers of complexity in the regulatory framework. **Our principal finding is that sequence-based *prediction* of Select Agent properties is not feasible, now or in the foreseeable future; any dedicated research effort *solely* for this purpose is likely to have *only* negative consequences.**

When the committee's report was in the final stages of completion, the White House issued on July 2, 2010, a new Executive Order, "Optimizing the Security of Biological Select Agents and Toxins in the United States." Although the committee did not have time to consider fully the implications of this Executive Order, it notes that several issues are particularly relevant to this report; these are briefly discussed in Box 1.2.

1

The Select Agent Regulations

CHARGE TO THE COMMITTEE

In 2006, the National Science Advisory Board for Biosecurity (NSABB) released a report, *Addressing Biosecurity Concerns Related to the Synthesis of Select Agents* (NSABB 2006)," which considered the impact of synthetic biology and DNA synthesis technology on biosecurity and the current Select Agents Regulations (SAR). The principal concerns it addressed were that:

- DNA synthesis technology is rapidly diminishing barriers to acquisition of pathogens, because an increasing variety of organisms may be instantiated by whole genome synthesis, rather than by transfer of samples of existing organism stocks or cultures;
- Natural variation and intentional genetic modification blurs the boundaries around any discrete list based on taxonomic names; and
- Synthetic biology may enable the accidental or deliberate creation of entirely novel pathogens unrelated to current ones.

The NSABB made four recommendations: (1) that the Department of Health and Human Services (DHHS) and the U.S. Department of Agriculture (USDA) clarify and harmonize guidance to investigators and synthetic DNA companies regarding genetic elements and nucleic acids subject to the Select Agent Regulations; (2) Synthetic DNA companies establish uniform and standardized policies to screen orders; (3) repeal 18 U.S.C. 175c, because current scientific insight precludes meaningful definition of an agent based solely on sequence homology; and (4) that two further studies be initiated. The first study would focus on reconciling the current Select Agent controls with advances in synthetic genomics (i.e., a new synthetic means of pathogen accessibility, other than by transfer of existing stocks or wild isolates). The second proposed study became the origin of our committee. This recommendation proposed that:

> . . . a group of experts from the scientific community be assembled to determine if an alternative framework based on predicted features and properties encoded by nucleic acids, such as virulence or pathogenicity, can be developed and utilized in lieu of the current finite list of specific agents and taxonomic definitions . . . (NSABB 2006).

Thus, this study was initiated with the title of "Scientific Milestones for the Development of a Gene Sequence-Based Classification System for Oversight of Select Agents" on the basis of this recommendation. The committee was specifically charged with identifying:

> . . . the scientific advances that would be necessary to permit serious consideration of developing and implementing an oversight system for Select Agents that is based on predicted features and properties encoded by nucleic acids rather than a relatively static list of specific agents and taxonomic definitions (see Box 1.1 or Appendix A).

BOX 1.1
Scientific Milestones for the Development of a Gene Sequence-Based Classification System for Oversight of Select Agents

Statement of Task

NIH has requested the National Research Council to convene an ad hoc committee to identify the scientific advances that would be necessary to permit serious consideration of developing and implementing an oversight system for Select Agents that is based on predicted features and properties encoded by nucleic acids rather than a relatively static list of specific agents and taxonomic definitions. The committee is asked to address several questions:

- What would be the key scientific attributes of a predictive oversight system?
- What are the challenges in attempting to predict biological characteristics from sequence?
- Does the current state of the science of predicting function from sequence support a predictive oversight system at this time?
- If not, what are the scientific milestones that would need to be realized before a predictive oversight system might be feasible?
- In qualitative terms, what level of certainty would be needed about the ability to predict biological characteristics from sequence data in order to have confidence in a predictive oversight system?
- In what time frame might these milestones be realized? What kinds of studies are needed to achieve these milestones?

ORGANIZATION OF THE REPORT

In order to think about what an alternative framework for Select Agents might look like in an era of genomics and DNA synthesis, Chapter 1 starts by presenting the rationale for and status of the current system. This chapter also discusses the consequences of the Select Agent Regulations on the gene synthesis industry and the research community, and the criteria that are considered when Select Agent status is determined. (Appendices for this chapter provide information regarding the Select Agent Regulations and related guidelines.)

Chapter 2 presents the complexity of biology, the challenges in predicting function from sequence, and the special case of synthetic biology. (Appendices for this chapter contain detailed information regarding the challenges in predicting pathogenicity from sequence, and include examples of "virulence factors" from known pathogens.)

Chapter 3 explains how prediction differs from classification, outlines three threat scenarios presented by synthetic genomics, and describes how—with current technologies—a sequence based classification system might be created to address the challenges that synthetic biology and natural variation present to biosecurity.

Chapter 4 summarizes key findings and conclusions, and provides near term milestones and long term areas of research that would enable a sequence based classification framework for biosecurity and biosafety. (Appendix L for this chapter presents additional milestones and issues for consideration regarding the development of a gene sequence based classification system.)

CONTEXT OF THE SELECT AGENT REGULATIONS

During the committee's deliberations, the issues of biological weapons, biosecurity, and biosafety were each considered relevant to addressing the charge. The Select Agent Regulations, the *NIH Guidelines*, and the implementing legislation for the BWC define conditions for the legitimate, safe, and secure use of biological agents that have the potential to be used for great harm. It is important that research be conducted on microbes including those that may cause disease; biosafety rules are needed to protect the environment and the health of clinicians, researchers and the public and are therefore a necessity—even in the absence of terrorism or other security threats. As will be discussed, the biosafety system that has been in place for decades provides a framework of oversight for legitimate use of pathogenic microorganisms. The BWC, and implementing legislation,[1] is concerned with illegitimate use of biological agents (i.e., as weapons). **The Select Agent Regulations identify agents most likely**

[1]The 1989 Biological Weapons Act is the implementing legislation for the BWC. Title 18 US, Chapter 10, §175 established criminal penalties related to the development, manufacture, transfer or possession of a biological agent, toxin, or delivery system for use as a weapon; the BWC has never attempted to list BW agents.

BOX 1.2
Presidential Executive Order

When the committee's report was in the final stages of completion, the White House issued on July 2, 2010, a new Executive Order, "Optimizing the Security of Biological Select Agents and Toxins in the United States" (see Appendix M). Although the committee did not have time to consider fully the implications of this Executive Order, it notes that several issues are particularly relevant to this report:

1) *Sec. 4. Risk-based Tiering of the Select Agent List.*
"Tiering and potential reduction of the Select Agent List: HHS and USDA will, through their current biennial process of reviewing the Select Agent List, tier the existing list based upon the risk posed by the pathogen or toxin in enabling a mass casualty incident through deliberate misuse. For those pathogens and toxins in the highest risk tier, HHS and USDA will evaluate options for the targeted application of physical security and personnel reliability measures in a manner commensurate to risk. HHS and USDA will also consider reducing the number of agents and toxins on the Select Agent List."

This is in good agreement with near-term milestone d (see Chapter 4), which discusses stratification or reduction of the Select Agent list. Prioritizing the Select Agent list based on risk would make any sequence-based approach to oversight more feasible.

2) *Sec. 7. Implementation. (a) Establishment, Operation, and Functions of the Federal Experts Security Advisory Panel. "A panel of Federal security and scientific experts will serve as the principal security advisory body to the SAP. The Panel will advise the SAP on a range of topics, including considerations in the tiering and/or reduction of the Select Agent List, best practices regarding physical security and personnel reliability that should be considered in the revision of the SAR and related Rules and guidance, and other topics as determined by HHS and USDA. The Department of Homeland Security will chair a sub-Group of the Panel that will advise the SAP on recommended physical security practices for high-risk pathogens and toxins. In addition, the EO directs the National Science Advisory Board for Biosecurity to serve as a source for external advice and input on SAP/SAR policies and practices."*

As discussed in the present report, (Chapter 3; Appendix L, near-term milestone f) a classification or yellow flag system would depend on the technical expertise of Scientific Workgroups and Advisory Panels. The Federal Experts Security Advisory Panel described in the EO could, if appropriately staffed with knowledgeable experts, be compatible with such scientific workgroups and advisory panels. For instance, the DHS sub-group is consistent with "Security Advisors" presented in Fig 4.1, 2b. It appears that the Federal Experts Security Advisory Panel could also include a sub group of "Scientific Advisors" as presented in Fig 4.1, 3b. (However, the FESAP may not be appropriate to perform the function of the yellow flag "Scientific Advisors," shown in Fig 4.1, 9b.)

to be used illegitimately, as weapons, restrict access to these microorganisms and their toxins to certain individuals, and specify conditions under which legitimate research use may occur. (See Appendix F for relevant legislation, regulation and guidelines.)

Biological Weapons

Pathogens have a long history of being investigated as potential offensive weapons for military purposes. It is well documented that the United States, the former Soviet Union, Great Britain, France, Germany and other countries maintained offensive biological warfare programs until signing of the Biological Weapons Convention of 1972.[2] In at least one case, that of the former Soviet Union, clandestine research and development continued well beyond this date and it was only in 1992, following the collapse of the Soviet Union, that Russia fully renounced biological weapons and opened some of its facilities to inspection by the international community. Other clandestine research is still the subject of considerable speculation but not unequivocally documented (Commission on the Prevention of Weapons of Mass Destruction Proliferation and Terrorism 2008).

Nations party to the BWC agreed to destroy or divert to peaceful purposes any existing weaponized biological agents or delivery systems within nine months of signing the convention. Violation of the terms of the treaty are not directly actionable by any oversight force; rather signatories are responsible for implementation through national means, which are disclosed on a regular basis.[3]

---

[2]The earliest global prohibition against the use of biological weapons is the 1925 Geneva Protocol. This treaty was augmented via the Convention on the Prohibition of the Development, Production and Stockpiling of Bacteriological (Biological) and Toxin Weapons and on Their Destruction (BWC), which was signed at London, Moscow, and Washington on 10 April, 1972, and entered into force on 26 March, 1975. Article I of the BWC states that:

"Each State Party to this Convention undertakes never in any circumstances to develop, produce, stockpile or otherwise acquire or retain:

(1) Microbial or other biological agents, or toxins whatever their origin or method of production, of types and in quantities that have no justification for prophylactic, protective or other peaceful purposes;

(2) Weapons, equipment or means of delivery designed to use such agents or toxins for hostile purposes or in armed conflict."

[3]"The United States Congress passed the Biological Weapons Anti-Terrorism Act of 1989 (Public Law 101-298, May 22, 1990), which established penalties for violating the Convention's prohibitions, unless "(1) such biological agent, toxin, or delivery system is for a prophylactic, protective, or other peaceful purpose; and (2) such biological agent, toxin, or delivery system, is of a type and quantity reasonable for that purpose." In keeping with the treaty, the legislation focused on the *purpose* for which agents or toxins were possessed, rather than the agents themselves. The law authorizes the government to apply for a warrant to seize any biological agent, toxin, or delivery system that has no apparent justification for peaceful purposes, but prosecution under the law

Nevertheless, concern about the use of biological weapons persists, as noted by the *Commission on the Prevention of WMD Proliferation and Terrorism* co-chaired by former Senators Bob Graham and Jim Talent. The 2008 report, *The World at Risk* (Commission on the Prevention of Weapons of Mass Destruction Proliferation and Terrorism 2008),[4] stated that "Unless the world community acts decisively and with great urgency, it is more likely than not that a weapon of mass destruction will be used in a terrorist attack somewhere in the world by the end of 2013."[5] More recently, the White House National Security Council issued "*The National Strategy for Countering Biological Threats* (National Security Council 2009)" at the December 2009 BWC meeting in Geneva. This report again addresses concerns that the life sciences might be misused.

Biosafety (and Categorization of Microorganisms)

In the United States, the Select Agent Regulations and other current legal mechanisms to control development, stockpiling, access, and use of specific biological agents had their roots in public health laboratory practices. In 1974, the Centers for Disease Control and Prevention (CDC) sought to limit the occurrence of laboratory-acquired infections by issuing guidance in the form of lists of microorganisms entitled, "Classification of Etiologic Agents on the Basis of Hazard" (CDC 1974). While adherence to safe practices was strongly encouraged, no federal requirements were imposed. However, these guidelines played an essential role in the development of the original *National Institutes of Health (NIH) Guidelines for Research Involving Recombinant DNA Molecules* issued in 1976[6] (NIH 1976). The same concerns for public health and safety led to promulgation of rules governing the packaging, labeling and transport of infectious agents shipped in interstate commerce (DHHS 1979). First published in 1984 by the CDC and National Institutes of Health (NIH), the document, "Biosafety in Microbiological and Biomedical Laboratories" (BMBL) defines the principles of biosafety for research and clinical laboratories around the globe. The BMBL describes safe handling practices and identifies four levels

---

would require the government to prove that an individual did not have peaceful intentions (Atlas 1999)" (NRC 2009b).

[4]See also, *The Clock is Ticking: A Progress Report on America's Preparedness* (Commission on the Prevention of Weapons of Mass Destruction Proliferation and Terrorism 2009).

[5]The validity of this statement is much debated. See for instance, "Does Threat Reduction Require Threat Inflation?" by Micheal Krepon and "Biological threats: A matter of balance," from the *Bulletin of Atomic Scientists*.

[6]The Recombinant DNA Advisory Committee (RAC) advised the NIH in the development of the *NIH Guidelines for Research Involving Recombinant DNA Molecules*, which have become the standard of safe scientific practice in the use of recombinant DNA. Institutional Biosafety Committees (IBCs), which are mandated by the *NIH Guidelines*, are charged with reviewing research involving recombinant DNA, although many IBCs have chosen to review other forms of research that involve potential biohazards—including research involving Biological Select Agent and Toxins (BSATs). Institutions are required to register their IBCs with NIH's Office of Biotechnology Activities.

BOX 1.3
The Nuclear Paradigm

Although biological materials have the potential for misuse and could be developed and employed as Weapons of Mass Destruction, it should not be assumed that biological agents are similar to nuclear WMD threats. Biological agents have unique attributes that are considered in order to develop an effective oversight strategy.

Characteristics of Fissile Materials and Pathogens

Fissile Materials	Pathogenic Microorganisms
• Do not exist in nature in readily concentrated form appropriate for weapons	• Generally found in nature and often widely distributed globally
• Non-living	• Living, replicative
• Difficult and costly to produce	• Easy and cheap to produce
• Not diverse: plutonium and highly enriched uranium are the only fissile materials used in nuclear weapons.	• Highly diverse
• Can be inventoried and tracked in a quantitative manner	• Vials containing pathogen cultures may be inventoried and tracked, however, because pathogens reproduce, exact pathogen quantification is unreliable.
• Can be detected at a distance from the emission of ionizing radiation	• Cannot be detected at a distance with available technologies
• Weapons-grade fissile materials are stored at a limited number of military sites	• Pathogens are present in many types of facilities (e.g. hospitals and schools) and at multiple locations within a facility
• Few non-military applications (such as research reactors, thermoelectric generators and production of radioisotopes).	• Many legitimate applications in biomedical research and the pharmaceutical biotechnology industry.

(Modified from Tucker 2003)

of containment based on risk criteria of infectivity, transmissibility, severity of illness, and the nature of work being performed. As a guideline, the BMBL maintains the flexibility needed to evolve along with the pathogens it describes. Though not a regulation, many choose to follow the BMBL's recommendations. Moreover, Select Agent Regulations part 73.12(c)(1) requires such consider-

BOX 1.4
NIH Guidelines—Risk Groups

Although not specifically targeted to Select Agents, the *NIH Guidelines for Research Involving Recombinant DNA Molecules* (*NIH Guidelines*) specify biosafety and containment practices for constructing and handling recombinant DNA. Agents are classified into four risk groups (RG) based upon their potential hazard. Select Agents are found in categorizes RG2, RG3, or RG4.

Risk Group 1 (RG1): Agents that are not associated with disease in healthy adult humans.

Risk Group 2 (RG2): Agents that are associated with human disease which is rarely serious and for which preventive or therapeutic interventions are often available.

Risk Group 3 (RG3): Agents that are associated with serious or lethal human disease for which preventive or therapeutic interventions may be available (high individual risk but low community risk)

Risk Group 4 (RG4): Agents that are likely to cause serious or lethal human disease for which preventive or therapeutic interventions are not usually available (high individual risk and high community risk)

The *NIH Guidelines* are normally enforced locally by Institutional Biosafety Committees (IBC), and each institution performing research with Select Agents usually have a dedicated Biosafety Officer who is responsible for ensuring regulatory compliance, reporting problems to the IBC, and serving as a technical resource for questions regarding biosafety. If Select Agents are present, the Biosafety Officer often serves as the first line of oversight in the management of research, transport and handling of Select Agents. A Responsible Official (RO) is designated for each entity holding Select Agents, and the Biosafety Officer may serve as the RO, or will work in close association with the RO.

ation as a result of a contractual requirement or best practices; thus in this case, the BMBL can be interpreted as having the force of law.

The BMBL, like the *NIH Guidelines*, stipulates four biosafety levels (BSL-1, -2, -3, -4) (Box 1.5). Increasingly stringent safety and containment criteria govern each biosafety level so that the pathogens with high mortality rates and no known treatment or prevention are designated BSL-4 agents. All BSL-4 agents have been designated as Select Agents; however, not all Select Agents require BSL-4 containment.[7]

The CDC conducted an assessment of the potential for biological agents to impact public health (Rotz, Khan et al. 2002). Criteria for classification included potential impact on public health, dissemination potential, public

[7] Anthrax, considered among the greatest security threats, requires only BSL-2/3 containment for biosafety, according to the BMBL.

BOX 1.5
BMBL Biosafety Levels

Biosafety Level 1	"BSL-1 is suitable for work involving well-characterized agents not known to consistently cause disease in immunocompetent adult humans, and present minimal potential hazard to laboratory personnel and the environment."
Biosafety Level 2	"BSL-2 is suitable for work involving agents that pose moderate hazards to personnel and the environment."
Biosafety Level 3	"BSL-3 is applicable to clinical, diagnostic, teaching, research, or productions facilities where work is performed with indigenous or exotic agents that may cause serious or potentially lethal disease through inhalation route exposure."
Biosafety Level 4	"BSL-4 is required for work with dangerous and exotic agents that pose a high individual risk of life-threatening disease, aerosol transmission, or related agent with unknown risk of transmission. Agents with a close or identical antigenic relationship to agents requiring BSL-4 containment must be handled at this level until sufficient data are obtained either to confirm continued work at this level or redesignate the level."

(CDC/NIH 2007)

perception, and special public health preparedness needs such as stockpile requirements, enhanced surveillance or diagnostic needs required to mitigate harm or respond following an attack. Public perception in this case refers to the ability to engender widespread panic or concern about the safety of products including food, thus precipitating a major impact even if an event does not cause direct or significant harm to human health. Individual pathogens were classified as Category A, B, or C, with Category A agents having the greatest potential to cause widespread casualties and requiring the largest public health preparedness efforts. Category B agents were also deemed to have a potential for large-scale dissemination, although generally with documented lower rates of illness and death. Agents not currently recognized as significant bioterrorism risks, but with attributes that might enable their future weaponization, were classified as Category C agents (Box 1.6).

The National Institutes of Health (NIH), National Institute for Allergy and Infectious Diseases (NIAID) has also developed a classification of pathogens using a Category A, B, and C system; however, this system is used to set research priorities and the criteria for classification was different. The NIH criteria stressed ease of dissemination, associated mortality rates following infection, potential for public health impact, social disruption, and required special action

BOX 1.6
CDC Bioterrorism Agents / Diseases by Category

Criteria	Category A	Category B	Category C[a]
Ease of dissemination	Easily	Moderately easy	Variable[a]
Mortality rates	High	Low	Variable[a]
Morbidity rates	Variable	Moderate	Variable[a]
Potential for social disruption	Possible	Variable	Variable[a]
Action for public health preparedness	Requires special actions	Specific enhancements of CDC's diagnostic capacity and enhanced disease surveillance	Variable[a]

[a]Category C considers emerging pathogens that could be engineered for mass dissemination in the future because of availability, ease of production and dissemination, and potential for high morbidity and mortality rates and major health impact.

SOURCE: Modified from Rotz, Khan et al. 2002).

for public health preparedness. A larger universe of pathogens was included in the NIH assessment such that some agents appear on the NIH list that were not captured on either the CDC classification or Select Agents list.

Biosecurity and the Select Agent Regulations

Biosafety and biosecurity[8] are related and complementary concepts; however, there are important distinctions. The fifth edition of the BMBL defines

[8]As discussed in NRC report 2009, "[i]t should be noted that the use of the term "biosecurity" presents a number of difficulties. At its most basic, the term does not exist in some languages, or is identical with "biosafety"; French, German, Russian, and Chinese are all examples of this immediate practical problem. Even more serious, the term is already used to refer to several other major international issues. For example, to many "biosecurity" refers to the obligations undertaken by states adhering to the Convention on Biodiversity and particularly the Cartagena Protocol on Biosafety, which is intended to protect biological diversity from the potential risks posed by living modified organisms resulting from modern biotechnology. (Further information on the Convention may be found at <http://www.cbd.int/convention/> and on the Protocol at <http://www.cbd.int/biosafety/>.) "Biosecurity" has also been narrowly applied to efforts to increase the security of dangerous pathogens, either in the laboratory or in dedicated collections; guidelines from both the World Health Organization (WHO 2004) and the Organization for Economic Cooperation and Development (OECD 2007) use this more restricted meaning of the term. In an agricultural context, the

biosafety programs as those that "reduce or eliminate exposure of individuals and the environment to potentially hazardous biological agents," while the "objective of biosecurity is to prevent loss, theft or misuse of microorganisms, biological materials, and research-related information" (CDC/NIH 2007). Biosecurity rose to public prominence in 1995, when three vials of an inactivated form of the organism that causes plague (*Yersinia pestis*) were illegally obtained by an alleged white supremacist (NRC 2009b). The perpetrator was charged with mail fraud as it was not a crime to possess these materials. This incident highlighted the need for a fundamental change in regulation of biological agents, which stimulated congressional interest culminating in the passage of legislation entitled "The Antiterrorism and Effective Death Penalty Act of 1996" (P.L. 104-132). This Act made it a federal crime to use or threaten to use a weapon of mass destruction including biological weapons and "directs the Secretary of Health and Human Services to promulgate regulations identifying biological agents that pose a potential threat to public health and safety and governing their intentional or inadvertent transfer." Accordingly, CDC promulgated regulations for facilities transferring or receiving select infectious agents and toxins as defined by the Secretary of Health and Human Services.

However, it wasn't until *Bacillus anthracis* was sent through the U.S. Postal Service in October 2001, resulting in five deaths, that a broader system of controls on the possession, use and transfer of Select Agents was established, carrying with it severe criminal penalties, including imprisonment and fines. The National Select Agents Registry Program (Select Agent Program) was formally established to execute provisions of the USA PATRIOT Act and the Public Health Security and Bioterrorism Preparedness and Response Act of 2002 regarding biological agents. Biological Select Agents and Toxins (BSAT) are defined by DHHS and the USDA as pathogens or biological toxins that have the "potential to pose a severe threat to public health and safety."[9]

The USA PATRIOT Act (Public Law 107-56) made it an offense for a person to knowingly possess any biological agent, toxin, or delivery system of a type or in a quantity that, under the circumstances, is not reasonably justified by prophylactic, protective, bona fide research, or other peaceful purpose. The Act also prohibited the possession or transfer of Select Agents by "restricted persons." A restricted person is defined by the Act (18 U.S.C. 175b) as an individual who:

- is under indictment for a crime punishable by imprisonment for a term exceeding 1 year;

---

term refers to efforts to exclude the introduction of plant or animal pathogens. (See NRC 2009a:8-9 for a discussion of this and other issues related to terminology.) Earlier NRC reports (2004ab, 2006, 2009ab) confine the use of "biosecurity" to policies and practices to reduce the risk that the knowledge, tools, and techniques resulting from research would be used for malevolent purposes."

[9]Both the 1996 and the 2002 legislation mandated a list of regulated agents, subject to biennial review and revision, or as needed.

- has been convicted in any court of a crime punishable by imprisonment for a term exceeding 1 year;
- is a fugitive from justice;
- is an unlawful user of any controlled substance as defined in section 102 of the Controlled Substances Act (21 U.S.C. 802);
- is an alien illegally or unlawfully in the United States;
- has been adjudicated as a mental defective or committed to any mental institution;
- is an alien (other than an alien lawfully admitted for permanent residence) who is a national of a country that has repeatedly provided support for acts of international terrorism; or
- has been discharged from the Armed Services of the United States under dishonorable conditions.

By prohibiting certain individuals from having access to Select Agents based upon criteria such as having committed a felony, convicted of illegal drug use, engaged in terrorist activities, or a history of mental illness, the Act addresses concepts related to the reliability of personnel in the research community.

The Public Health Security and Bioterrorism Preparedness and Response Act of 2002 (P.L. 107-188) requires the DHHS to regulate biological agents and toxins that have the potential to cause a severe threat to public health and safety and the U.S. Department of Agriculture to regulate biological agents or toxins that have the potential to pose a severe threat to animal or plant health or animal or plant products. These rules put in place a system of safeguards that were intended to allow scientists to conduct research without undue burden, while reducing "the risk of illicit access to these dangerous human pathogens." The statute (Part B of Section 511) directs the Secretaries of DHHS and USDA to establish and maintain lists of biological agents and toxins, to be reviewed at least every two years. It also specifies that in developing these lists, consideration should be given to: (I) the effect on human health of exposure to the agent or toxin; (II) the degree of contagiousness of the agent or toxin and the methods by which the agent or toxin is transferred to humans; (III) the availability and effectiveness of pharmacotherapies and immunizations to treat and prevent any illness resulting from infection by the agent or toxin; and (IV) any other criteria, including the needs of children and other vulnerable populations, that the Secretary considers appropriate; and (ii) consult with appropriate Federal departments and agencies and with scientific experts representing appropriate professional groups, including groups with pediatric expertise. The final rule implementing this legislation was published in 2005 and seeks to harmonize requirements of the two oversight bodies, the CDC acting on behalf of the Secretary of the DHHS and the Animal, Plant Health Inspection Service (APHIS), acting on behalf of the Secretary the USDA.

The current list of approximately 80 Select Agents and Toxins is shown in Appendix C. At present, the Select Agent list does not prioritize those biological agents based on security risk.[10] The regulations require that the institution's security plan, designed in accordance with site-specific risk assessment, provide graded protection in accordance with the risk of the Select Agent or Toxin, given its intended use. The CDC and APHIS administer the Select Agent Program, which ensures that those registered to possess, use or transfer Select Agents are in compliance with the Select Agent Regulations.

The Department of Homeland Security (DHS) was established on November 25, 2002, by Congress under provisions of the Homeland Security Act of 2002. It was intended to consolidate U.S. executive branch organizations related to "homeland security" into a single cabinet agency. The DHS conducts a biennial Biological Threat Risk Assessment (BTRA), the specifics of which are classified (The White House 2004). Assessments are currently based on lists of known pathogens, consideration of traditional scenarios and agents, and include specific traits such as antibiotic resistance. Techniques used for analysis in the BTRA continue to evolve; it is moving towards a systems biology approach that would be applied to characterize risk-associated attributes based on the modern tools of molecular biology and to consider scenarios with constructed novel organisms with defined pathogenic characteristics (NBACC 2009). The DHS believes a systems biology framework built upon a foundation of knowledge of the biological characteristics of traditional agents may provide a robust assessment of potential advanced threats (see also Box 1.7).

The U.S. government also created the National Science Advisory Board for Biosecurity (NSABB) in 2005 based in the Office of Science Policy of the NIH, to provide advice, guidance, and leadership regarding biosecurity oversight of dual use research. Dual-use research in this context is defined as biological research with legitimate scientific purpose that may be misused to pose a biologic threat to public health and/or national security.

CURRENT STATUS OF THE SELECT AGENT PROGRAM

The Select Agent Program—Beyond Biosafety

While there is considerable overlap in good laboratory practices directed toward both ends, the Select Agent list highlights the distinction between biosafety and biosecurity. Biosafety refers to mitigation of the risk of pathogens or toxins escaping containment and causing illness in laboratory workers or the

[10] Several groups have recently recommended prioritization of the Select Agents list, including the NRC BSAT committee, the Interagency Working Group on Strengthening Biosecurity, as well as legislation introduced by Senators Lieberman and Collins (S.1649-WMD Prevention and Preparedness Act of 2009), which calls for a "tiered" approach.

BOX 1.7
Biological Threat Risk Assessment

The Department of Homeland Security's Biological Threat Risk Assessment (BTRA) of 2006 is classified; the specific criteria used to perform the risk assessment are therefore unavailable. BTRA methodology, however, does not allow for categorical criteria:

"DHS chose to assess threat by ranking bioagents because government stakeholders had advised DHS that they "expected the primary assessments to be in the form of risk-prioritized groups of biological threat agents" (DHS 2006). Although a terrorist's choice of agent is just one step in a sequence of events leading to a potential attack, for practical purposes the BTRA of 2006 evaluates each agent separately. A probability is computed for each scenario involving that agent. Risk is then calculated as the product of these probabilities and the associated consequences. The overall risk assocaited with each agent is the integrated risk distribution over all possible scenarios involving that agent (NRC 2008a)."

"The process that produced the estimates in the BTRA of 2006 consists of two loosely coupled analyses: (1) a probabilistic risk analysis event-tree evaluation and (2) a consequence analysis." (NRC 2008)

The BTRA was updated in 2008 and 2010. Risk is evaluated on an agent-by-agent basis with an objective to stratify the analysis by different elements such as exposure routes, biological agent targets, acquisition, production, method of dissemination, as well as likelihood of deployment (e.g. by specific terrorist groups). The model has multiple nodes aimed to maximize oppertunities to understand what is and isn't known at varying degrees of granulatity (Pesenti 2009).

general public. Biosecurity is directed toward minimizing the possibility that such pathogens or toxins will be misused, stolen, diverted, or intentionally released (Working Group on Strengthening the Biosecurity of the United States 2009). Thus, the Select Agent Regulations are designed to prevent unauthorized access, theft, loss, or release of Select Agents and Toxins (Appendix C).

The Select Agent Regulations require entities to provide graded protection to Select Agents and Toxins including limiting access to Select Agents and Toxins according to a site-specific security plan required for each institution registered with CDC or APHIS as authorized to possess and use such biological materials.[11] Entities using strictly plant or animal pathogens report to APHIS; entities using human or zoonotic pathogens report to CDC. Institutions must submit the names of individuals who will be allowed access to Select Agents for a background check—termed a security risk assessment (SRA)—that is con-

[11]CDC and APHIS report that as of September 2009 there were 388 enteties authorized to work with Select Agents ad Toxins (NRC 2009b).

ducted by the Federal Bureau of Investigation's Criminal Justice Information Services Division every five years. Registered institutions must also provide a list of agents in use or intended for use, and must report when any changes in use take place. Perhaps the most controversial element of the reporting requirements is the need to keep a running inventory of stock, which is a challenge in the case of living, reproducing organisms.[12] The Select Agent Regulations requires registered entities to maintain records pertaining to Select Agent access, long term storage inventories, transfers, training, etc. for a period of three years. Entities are required to maintain current and accurate records, and implement a system that ensures that all records and data bases created under the rule are accurate, have controlled access, and that their authenticity may be verified.

In addition, entities must conduct annual inspections for each laboratory where Select Agents or Toxins are stored or used in order to determine compliance with the Select Agent Regulations, the results of which must be documented, and any deficiencies identified must be corrected. Failure to meet these requirements may result in criminal penalties of fines and up to ten years imprisonment. Thus, the Select Agent Regulations can be reasonably viewed as an instrument of law enforcement to facilitate attribution[13] and prosecution in the event of domestic use or, deliberate or inadvertent possession of potential biological weapons.

The Select Agent Program—Focus on *Known* Biothreat Agents

The Select Agent Program was devised to establish controls for *known* biological agents and toxins that have the potential to pose a severe threat to public health and safety. Newly emerging pathogens are not given Select Agent

---

[12]This issue was addressed in Recommendation 4 of the NRC report, Responsible Research with Biological Select Agents and Toxins. —"Because biological agents have an ability to replicate, accountability is best achieved by controlling access to archived stocks and working materials. Requirements for counting the number of vials or other such measures of the quantity of biological Select Agents (other than when an agent is transported from one laboratory site to another) should not be employed because they are both unreliable and counter-productive, yielding a false sense of security. A registered entity should record the identity of all biological Select Agents and toxins within that entity, where such materials are stored, who has access and when that access is available, and the intended use(s) of the materials" (NRC 2009b).

[13]Microbial forensics plays an important in attribution efforts. Microbial forensics, also biofo-rensics, is a relatively new scientific discipline that draws from other science disciplines including, genomics, microbiology and plant pathology (Breeze, Budowle, et al., Eds. (2005). *Microbial Forensics*, Elsevier Academic Press. Microbial forensics is dedicated to analyzing microbial activity as evidence for attribution purposes and/or back tracking. Microbial forensics procedures support 'decision taking' at biosecurity levels, follows strict chain of custody of specimens and demands a rigorous (accredited) and unbiased performance. Therefore, microbial forensics includes the complete range of forensic evidence analysis from microorganisms, to associated evidence materials found at the field of the suspected outbreak or crime scene (Breeze, Budowle, et al., Eds. (2005). *Microbial Forensics*, Elsevier Academic Press.

status until they have been assessed.[14] When a novel agent emerges, it is named. Research is initiated to study its mechanism of action, the potential threat it presents, and its susceptibility to countermeasures. After knowledge is gained, an agent may be added to the list. Pathogens and toxins may be considered for inclusion on the Select Agent list at the discretion of the Secretary of DHHS and often at the suggestion of the Intragovernmental Select Agents and Toxins Technical Advisory Committee (ISATTAC), which is charged with making such recommendations. This process requires consultation with appropriate federal agencies and with scientific experts representing appropriate professional groups.

Importantly, this model allows some ability to consider the impact that the Select Agent Regulations may have on legitimate research before an agent is added to the list. Significant constraints on research may have negative consequences for public health and security by impeding a vigorous research enterprise response, which provides the foundation for the creation of new diagnostics, vaccines and therapeutics. This paradox provided early justification for not including the causative agent of Severe Acute Respiratory Syndrome (SARS-Corona virus) on the Select Agent list in early 2004, leading to containment of the epidemic by unparalleled international communication and collaboration.

Therefore, it is the committee's view that the Select Agent Regulations are necessarily backward-looking and based on a list of known agents. It is clear, however, that there are unforeseen natural threats, as well as the potential to develop novel pathogens that are not in the realm of contemporary classification. Such unknown, unnamed pathogens are not yet Select Agents; these novel agents present a particularly challenging issue of direct concern to this committee.

Although the Select Agent Regulations do not focus on novel agents, the Select Agent Program has, since its inception, attempted to address the possibility that genetic material derived from a Select Agent might be used to construct a biological threat agent. Language from the *NIH Guidelines for Research Involving Recombinant DNA Molecules* was adopted in the Select Agent Rule to address this concern:

> Select agent means a microorganism (virus, bacterium, fungus, rickettsia) or toxin listed in Appendix A of this part.
>
> The term also includes:
>
> (1) Genetically modified microorganisms or genetic elements from organisms on Appendix A of this part, shown to produce or encode for a factor associated with a disease, and

[14]As will be discussd, the criteria for inclusion of any biological agent or toxin onto the Select Agent list is provided in Public Health Security and Bioterrorism Preparedness and Response Act of 2002.

> (2) Genetically modified microorganisms or genetic elements that contain nucleic acid sequences coding for any of the toxins on Appendix A of this part, or their toxic submits (DHHS 1996).

It is important to note that this language is absolute and refers to direct evidence or experimental knowledge of the ability of the proteins encoded by the nucleic acids to cause biological damage. The concept that nucleic acid sequence might serve as a *predictor* of function of the encoded product was not addressed. Contemporaneous versions of the *NIH Guidelines* attempted to deal with the possibility that partial genomes could confer pathogenicity or toxicity by inclusion of a proxy, stating that transfer of more than "two-thirds of the genome" of a eukaryotic virus would likely necessitate a higher level of containment.[15,16]

We currently live in a scientific environment in which constructing a known gene(s) or modifying a microorganism may be achieved largely through synthetic means (See Cello, Paul et al. 2002; Gibson, Glass et al. 2010). In this regard, the CDC has issued guidance on the "Applicability of the Select Agent Regulations to Issues of Synthetic Genomics" (see Appendix E). Again, this guidance deals with known microorganisms, already designated as Select Agents. Its purpose is to clarify that Select Agent Regulations apply, even if the Select Agent is produced by synthetic means or has been genetically modified. The guidance does not attempt to address novel organisms that are not already designated as Select Agents. While this highlights the focus of the Select Agent Regulations on known threats, it also indicates the difficulty in setting clear boundaries around diverse organisms.

Unclear Boundaries

There is no taxonomic foundation for designation of a pathogen as a Select Agent. Even with the full genomic sequence of multiple strains of Select Agents, the genetic differences between a defined Select Agent and closely related non-Select Agents are sometimes blurred and may lead to questions and uncertainty by both the scientific community and those responsible for

---

[15]The *NIH Guidelines* are in the process of being amended to address this issue. The proposed revision was published in the April 22, 2010 Federal Register (http://oba.od.nih.gov/oba/RAC/2010-9258.pdf).

[16]The U.S. Patent and Trademark Office has struggled with a similar question since the early 90's. Large-scale sequencing efforts resulted in an enormous number of patent applications on human gene sequences. In order to meet the legal criterion for proof of utility, applicants provided computer-generated comparisons demonstrating the similarity or homology of a claimed sequence to another published sequence of known function. There were heated discussions about the validity of using an algorithm to predict how the product of sequence might behave in a biological system.

oversight of the Select Agent program (Casadevall and Relman 2010). A case in point is the stipulation in the 18 USC Part 1, Chapter 10, subsection 175 c. that established criminal penalties for the possession of biological material sharing 85 percent or more of the genome of variola virus, the cause of smallpox. It is unclear what "85 percent of the gene sequence" means. Does this refer to a fragment of 85 percent or more of the full-length variola virus genome? Or does it mean a full-length genome of 85 percent or greater sequence identity to variola? Moreover, several orthopoxviruses share approximately 96 percent amino identity with variola. The language is sufficiently problematic that the NSABB has recommended the repeal of the requirement because it may inadvertently criminalize work on other orthopoxviruses including vaccinia, the smallpox vaccine virus, which was surely not the intent of the legislation[17,18] (NSABB 2006). This example highlights the challenges faced in defining Select Agents based on sequences alone.

Gene Synthesis Industry

If genetic sequences do not suffice to define Select Agents, then a challenge arises in providing guidance to the approximately fifty companies worldwide that offer gene synthesis as a commercial service (Maurer, Fischer et al. 2009). With one exception,[19] all of the gene synthesis companies are private and none publish their financial results. Several manufacturers market their services through more than one distributor. The fact that the market is largely privately held and highly dispersed makes it difficult to gauge its size with any accuracy, but estimates place its value between $50 and $80 million in 2008. This represents over 50 million bases pairs of synthetic DNA, or approximately 75,000 genes. The cost of synthesis has declined rapidly, dropping by half every 18-24 months over the last ten years and will likely continue to decline for at least the next few years. Each of the top gene synthesis companies is currently capable of providing 1-2 million base pairs of synthetic DNA per month, and most are planning for rapid growth.

---

[17]In July 2008 the Department of Justice released a memorandum clarifying the scope of the definition of variola virus under the Intelligence Reform and Terrorism Prevention Act of 2004, Section 6906, making it a criminal offense to knowingly produce, engineer, synthesize, acquire transfer directly or indirectly receive, possess, import, export or use, or possess and threaten to use, variola virus. 18 USC 175c (a) (1) exempts work conducted by or under the authority of the Secretary of HHS. The Department of Justice stated that section 175c does not apply to all orthopoxviruses, but only to viruses that cause smallpox. However, the challenge remains—how can one tell from sequence if an orthopoxvirus will cause smallpox?

[18]This is still active statute; however, S.1649-WMD Prevention and Preparedness Act of 2009, was amended in mark up to include an amendment which codifies the WHO recommendations for the distribution, handling and synthesis of Variola Virus DNA and mandates regulations governing the distribution, synthesis and handling of variola virus DNA.

[19]GeneArt.

While gene synthesis provides many benefits to biomedical research, it also provides an alternate route of access to toxins and pathogenic organisms. Current government oversight of the DNA synthesis industry has not kept up with the pace of science in a number of important ways. First, existing regulations are difficult to interpret in the context of a series of DNA sequence orders. The Select Agent Regulations focus on species definitions of pathogens and do not define the boundaries between a pathogen and a similar sequence from a related species. (The Export Administration Regulations enforced by the U.S. Department of Commerce include phrases such as ". . . sequences associated with the pathogenicity of microorganisms . . ." that are vague and imprecise.) Second, the regulations do not address activities that have the potential to cause significant alarm in the public and the scientific community. For instance, two DNA fragments that comprise the genome of the Select Agent Omsk Hemorrhagic Fever virus are not covered by the Select Agent Regulations when separated, but would be covered if the two fragments are joined together using simple, widely available molecular tools.

Gene synthesis companies have been active in working with their respective governments to promote effective regulation of the technology. All gene synthesis companies must comply with their national laws regarding pathogen and toxin DNAs.[20] Most of the companies have customers in more than one country and must also comply with regulations that govern export from their country, as well as the regulations governing importation and possession of the sequences at the customer's location. Several different industry groups have formed to promote safe, effective regulation of gene synthesis technology.

Although the application of the Select Agent Regulations to cloned DNA could be interpreted in a number of different ways, the CDC has provided a guide to interpreting these rules for the gene synthesis companies ("Applicability of the Select Agent Regulations to Issues of Synthetic Genomics," see Appendix E). The guidance document defines the organisms whose genomes are covered and provides examples that clarify the application of these rules. The document does not define the species boundaries in terms of sequence similarity. The gene synthesis companies therefore pragmatically must define a number of critical screening parameters on their own. However, because they must rely on sequence alone, they are faced with significant questions: Which

[20]Gene synthesis companies based in the United States are affected by at least four regulations that cover the synthesis of pathogen and toxin genes; (1) the Select Agent Rules, (2) the Department of Commerce Export regulations, (3) the Biological Weapons Anti-Terrorism Act of 1989 and (4) a specific statute relating to smallpox. In addition, the World Health Organization places significant limits on the possession and end use of small pox genes, including a prohibition against the synthesis of any fragment of more than 500 base pairs. Currently, each of the approximately 40 companies worldwide that supply synthetic genes has developed an independent process to screen orders for sequences that might be covered by the laws and regulations of its own country and those of the customer's country.

genome should be used as a reference for the species? Is a sequence that is 99 percent identical to the reference genome covered? 98 percent? 90 percent?

DHHS has recently issued a document "Screening Framework Guidance for Synthetic Double-Stranded DNA Providers," which provides a more detailed methodology by which companies can screen DNA synthesis orders. These draft guidelines have been published in the Federal Register, and are currently available for public comment prior to issuance of a final set of guidelines. The guidance requests that companies identify and follow up on sequences (greater than 200bp) with homology *unique* to a Select Agent sequence. Thus, the guidance aims to define the boundaries between a Select Agent and a similar sequence from a related species. The guidance also discusses concerns that virulence genes from non-Select Agents could be used to produce biological threats:

> The U.S. Government acknowledges that there are synthetic nucleic acid sequences from non-Select Agents or Toxins that may pose a biosecurity concern. Synthetic nucleic acid providers may choose to investigate such sequences as part of their best practices. However, due to the complexity of determining pathogenicity and because research in this area is ongoing, a list of additional non-Select Agent or Toxin sequences or organisms to screen against would not be comprehensive and consequently are not provided by the U.S. Government in this guidance (DHHS 2009).

However, the guidance does encourage companies to "exercise their due diligence in the investigation of screening hits against non-Select Agents and Toxins that may raise a biosecurity concern" (DHHS 2009). While the motivation for this portion of the guidance is clear, the ability to implement the suggestion is not. Companies may find it challenging to recognize a sequence that poses a biosecurity concern, readily identify virulence genes, or determine the usability of genomic sequences in predicting potentially dangerous sequences in naturally occurring, genetically modified, or synthetically derived microorganisms.[21]

Impact of Select Agent Regulations on Research

There is a clear recognition that research on infectious disease agents, including Select Agents, is vital to public health and national security. There is also an acknowledgement among researchers that Select Agents should be regulated (NRC 2009b; Sutton 2009). However, for those scientists who choose to pursue this line of investigation, it is not always clear what organisms and which

[21]The draft DHHS screening guidelines were the topic of a January 11, 2010, meeting hosted by AAAS CSTSP. A summary of the "major themes of the meeting, including concerns and/or challenges highlighted, and recommendations proposed by individual attendees" is available online. Berger, K. M., W. Pinard, et al. (2009). Minimizing the Risks of Synthetic DNA: Scientists' Views on the U.S. Government's Guidance on Synthetic Genomics.

researchers must comply with the SAR. The regulations as currently written are open to interpretation, due to the unclear boundaries discussed above. For example, SARS-CoV is not currently a Select Agent, but may soon be designated as such. If this agent is added to the list, researchers will be required to register their SARS-CoV strains. But what does this mean? A SARS-CoV researcher is likely to have scores of related or derivative viruses, whose genomes are not identical to the 'original' SARS-CoV genome sequence.[22] The various strains may or may not be pathogenic, and may have been obtained from the wild or via genetic modification of the 'parent' virus. Keep in mind that a single nucleotide change may render the organism non-pathogenic, whereas multiple-changes elsewhere could have no effect on pathogenicity.[23] A researcher may need to register all of their various strains; or if none of the sequences match the 'original' SARS-CoV sequence, then the researcher may believe that registration is not required. This is a serious concern. A misinterpretation of the requirements could lead to a substantial burden in time and resources as an investigator complies with the Select Agent Regulations unnecessarily; on the other hand, non-compliance could lead to criminal prosecution. Thus, having a clear definition of what is and is not a Select Agent, is vitally important to the responsible scientist (See Sutton 2009 and Box 1.8).

The designation of certain infectious organisms and toxins as Select Agents with the potential to be used as bioweapons challenged both policy makers and the scientific community to understand better the pathogenic mechanisms of these microorganisms and toxins and to develop countermeasures to prevent, diagnose, and treat the effects of such agents. Paradoxically, the designation of these organisms and toxins as Select Agents put considerable burden on the scientific community to conduct this research while simultaneously adhering to costly and rigorous standards for security and accountability (Dias, Reyes-Gonzalez et al. 2010).

The Select Agent Program was not designed to impede research but there is much concern in the scientific community that the requirements have resulted in unintended negative consequences (NRC 2009b; Dias, Reyes-Gonzalez et al. 2010). Select Agent Program criteria would be most useful if based on genomic information and a rigorous biological foundation tempered by the realities of national and international security.

[22]Moreover, no reference sequences are currently provided for agents on the Select Agent list. In this regard, the agents are defined based on taxonomy and "chain of custody."

[23]As discussed in the Chapter 2, the effect that these sequence changes have on pathogenicity isn't known until experiments are done. It should also be noted that a nucleotide sequence may contain "silent" mutations, which alter the DNA sequence, without affecting the protein sequence. Therefore, there may be dissimilarity at the DNA sequence level and identity at the protein level.

BOX 1.8
Effect of Select Agent Regulations on Research

During deliberations, the committee noted that the Select Agent Regulations have potentially significant consequences for the scientists who work in this area and on the research that is done. The Select Agent Regulations are based in law and backed by criminal penalties. Implementation of the current Select Agent Regulations was therefore discussed in the context of what a "predictive oversight system" might look like and the "level of certainty" such a system would require. The committee was aware that several issues of concern have been noted by the scientific and security communities. Although the quantitative affect of the Select Agent Regulations on research is unknown, some of the concerns are briefly presented below:

- *Altered Research Direction*—Scientists may have redirected their research to the study of attenuated strains that are not classified as Select Agents (e.g. *Bacillus anthracis* Sterne strain) (CDC 2005). "In our view, the most likely explanation for the 10-fold discrepancy in the number of toxin- and capsule-related papers is that capsule-related research must be carried out within the SATL-associated regulations. If, in fact, these regulations are hindering capsule-related research, such hindrance has direct biodefense and preparedness implications, given that capsule components have been shown to be effective vaccines" (Casadevall and Relman 2010).
- *Considerable costs*—As noted by the NRC *Responsible Research with Biological Select Agents and Toxins*, "select agent laboratories have significant ongoing security and safety sustainment costs that far exceed the indirect costs that grantee institutions receive to cover the costs of facilities, maintenance, and operations." Thus, in the absence of continued federal funding, the institution must be willing and able to commit funds to meet this additional financial burden or instead deny their investigators such Select Agent research pursuits (NRC 2009b).
- *Inefficiency*—Funds awarded to principal investigators for Select Agent research do not go as far as they would be expected to in a non-Select Agent laboratory. This may be due to the additional manpower needed for intensive recordkeeping (lab

CRITERIA FOR SELECT AGENT DESIGNATION

How should a "Select Agent" be defined? The registry of Select Agents is a finite list that has grown in recent years as emerging pathogens are characterized and assessed. For example, public comments are now being requested as officials consider whether Severe, Acute Respiratory Syndrome (SARS) coronavirus and a recently recognized arena virus, Chapare virus, should be added to the Select Agent list. In some instances naturally occurring or derived attenuated strains of Select Agents have been removed from the list; however, despite legislation requiring periodic review of the Select Agent list, it is currently challenging to subtract from the list because there are no precise biological or policy access, equipment access, reagent use, etc.), and the extensive recurrent training necessary for Select Agent registration compliance. There is inherent down time during which individuals must wait to be cleared by the CDC and FBI to begin work in the Select Agent areas. As noted by Dias et al., "the most striking effect observed was not associated with individual authors or institutions; it was a loss of efficiency, with an approximate 2- to 5-fold increase in the cost of doing select agent research as measured by the number of research papers published per millions of U.S. research dollars awarded"(Dias, Reyes-Gonzalez et al. 2010).

- *Personnel Issues*—As noted by the NSABB, "Certain research facilities (notably federal) have instituted formal Personnel Reliability Programs (PRPs) to provide additional measures to help ensure that individuals with access to select agents meet additional standards of reliability . . . . "The promulgation of additional reliability measures could serve as a powerful disincentive to those who wish to and would responsibly conduct research on select agents because the most talented young researchers, those with many options for research paths, may be far more likely to enter fields with less onerous regulatory requirements. Thus, a burdensome national personnel reliability program may not only drive scientists from important select agent research, but also drive select agent research out of academia and potentially out of the U.S. into countries with less stringent regulations (NSABB 2009).
- *Collaborations* may be affected between Select Agent laboratories and non-Select Agent-certified collaborators and trade partners, especially overseas. The development of diagnostics and vaccines may require the sharing of samples, recombinant DNA, or toxins in quantities above Select Agent limits. Commercial partners may view the requirement of being Select Agent-registered as a disincentive to collaboration and development of essential biodefense-related preventives, therapeutics, and diagnostics. This concern has again been expressed in response to a proposal of adding SARS-CoV to the SA list, "The ASM believes the proposal to add the SARS-CoV virus to the HHS list of select agents and toxins deserves further deliberation . . . consider the impact on international collaboration and public health efforts."

criteria to do so. Nor are there precise, quantitative criteria for inclusion. Select Agent status has generally been conferred on an *ad hoc*, case-by-case basis, using a combination of the following considerations.[24]

[24]"[T]he Act requires the HHS Secretary to consider the following criteria in determining whether to list an agent or toxin: (1) The effect on human health of exposure to the agent or toxin; (2) the degree of contagiousness of the agent or toxin and the methods by which the agent or toxin is transferred to humans; (3) the availability and effectiveness of pharmacotherapies and immunizations to treat and prevent any illness resulting from infection by the agent or toxin; and (4) any other criteria, including the needs of children and other vulnerable populations, that the Secretary considers appropriate" (DHHS 2005).

- Virulence, pathogenicity, or toxicity of the organism; its potential to cause death or serious disease.
- Availability of treatments such as vaccines or drugs to control the consequences of a release or epidemic.
- Transmissibility of the organism; its potential to cause an uncontrolled epidemic.
- Ease of preparing the organism in sufficient quantity and stability for use as a bioterrorism agent; for example, the ability to prepare large quantities of stable microbial spores.
- Ease of disseminating the organism in a bioterrorism event to cause mass casualties, for example by aerosolization.
- Public perception of the organism; its potential to cause societal disruption by mass panic.
- Known research and development efforts on the organism by national bioweapons programs.

Thus, there are multiple factors that are considered before adding an agent to the Select Agent list. Simply possessing the capability of causing a significant threat to public health and safety does not meet the threshold for designation as a Select Agent. For example, seasonal influenza causes an average of 35,000 deaths annually in the United States, yet it is not considered a Select Agent (an effective vaccine is available). Historically, pathogens that were previously weaponized either by the United States or other countries are considered to be the greatest risk,[25] even when subsequent scientific findings suggest otherwise. Potential biological weapons threats are uniformly discussed (See, for example, Kortepeter and Parker 1999), as possessing attributes that would enhance their appeal to terrorists as a weapon. Since the 1950s these attributes include the ability to incapacitate affected individuals or cause highly lethal infections in a short period of time, lack of availability of preventive or therapeutic measures, ease of production, stability as an aerosol, and ability to be dispersed as small particles, all characteristics that could lead to significant loss of life, overwhelm the healthcare system, and cause social disruption and panic. For example, variola (the virus that causes smallpox) is perhaps the clearest and least controversial agent on the Select Agent list; smallpox virus has the potential to cause a catastrophic epidemic if it were released.

It also seems clear that the aim of the Select Agent Regulations is not to regulate *all* organisms that could be used by bioterrorists, even if the organism has already been used for bioterrorism. *Salmonella typhimurium* was the agent used in a 1984 incident in Oregon by followers of Bhagwan Shree Rajneesh, but

[25]Those pathogens most often considered as greatest threats included anthrax (*Bacillus anthracis*), botulinum toxin, tularemia (*Francisella tularensis*), plague (*Yersinia pestis*), and smallpox (variola virus), among others.

the resulting cases of food poisoning were on a scale similar to many normally occurring food contamination events and readily handled by the normal public health system.

One useful source for understanding the rationale for the current Select Agents list is the commentary on 42 CFR Part 73, the DHHS implementation of provisions of the Public Health Security and Bioterrorism Preparedness and Response Act of 2002 (DHHS 2002), describing some of the considerations that led to the Select Agents list that superseded the 1997 list in 42 CFR 72.6. For example, "viruses causing hantavirus pulmonary syndrome" were removed from the CFR 72.6 Select Agent list because they "are difficult to propagate and there is a lack of data establishing laboratory acquired infections." Yellow fever virus was removed because "there is a safe and effective vaccine." *Histoplasma capsulatum* and *Blastomyces* species were considered but not included because "they are difficult to cultivate and do not sporulate readily." Aflatoxins were not included because "the acute toxicity is too low to pose a significant mass casualty threat." The published discussion in the Federal Register does not include any examples in which DHHS considered public perception of the organism or known bioweapons development programs in its decision making. It explicitly lists "effect on human health" (pathogenicity), "contagiousness," "methods by which the agent or toxin is transferred to humans," (dissemination) and "the availability of pharmacotherapies and immunizations" as criteria, and the discussion includes more than one example in which ease of preparation of large quantities of the organism was considered. All in all, the decision to include a microorganism or a microbial product as a Select Agent is based on a variety of factors, including past medical experience, partial laboratory evidence, and historical precedent.

From this perspective it is worth considering which properties (or general criteria) can be predicted from genome sequence now, which ones might be predicted in the future, and which ones will never be predictable from genome sequence because they are not biological properties.

Non-Biological Criteria

As mentioned, non-biological information is considered when determining if an agent should be designated as a Select Agent or Toxin. Factors such as trade policy, the availability of therapeutics, natural prevalence of the microorganism, and historical use of an agent as a bioweapon can all effect whether or not a microorganism poses a threat to national security (Table 1.1). Because such factors are not inherent biological properties, they can never be determined by a biological agent's genome.

For example, USDA animal pathogens are included in the Select Agent list because the diseases they cause have significant economic and trade repercus-

TABLE 1.1 Prospects for *de Novo* Prediction of "Select Agent-ness" from Sequence

Property	Predictable Now?	Foreseeable Future?	Maybe Someday?	Never
Pathogenicity			X	
Transmissibility			X	
Available treatments			X	
Ease of preparation			X	
Ease of dissemination			X	
Public perception				X
Historical bioweapon				X
Economic impact				X
Natural prevalence				X

sions for U.S. Agriculture.[26] A few years ago U.S. international trade in beef was halted for approximately two years at a cost of billions of dollars to U.S. farmers due to the occurrence a single case of bovine spongiform encephalopathy (BSE)—commonly known as "mad-cow disease." This response was the result of international trade rules—long supported by the U.S.—that mandate that all animal product exports being halted in the event of a single case of foot and mouth disease or BSE (or other transboundary livestock and poultry diseases listed by the World Animal Health Association[27]). The concept of "increased virulence" is thus intriguing in regard to these non-zoonotic animal Select Agents; if the catastrophic response is triggered, as for BSE, by even one case, then the biological characteristic of virulence is not an issue. What could possibly be made worse? Even a totally synthetic foot and mouth virus designed for enhanced virulence would have no increased impact from the point of view of international trade rules. Because it is not U.S. policy to vaccinate animals to control the disease, FMD is already the most infectious animal virus known (a single pig is estimated to produce 100 billion cattle infectious doses per day and there are approximately 24 million pigs in Iowa) and the international trade consequences would be the same as a natural FMD virus. Thus, these agents are designated as Select Agents not because they pose a threat to human health, or even animal health. Rather, these agents pose a threat to national security, and are designated as Select Agents because of potential economic consequences, international trade agreements, and vaccination policy.

Thus, criteria that are considered in designating a microorganism as a "Select Agent" include biological and non-biological data. While it is not cur-

[26]The international trade rules for animal agriculture are set by the International Office of Epizootics in Paris (recently retitled the World Animal Health Association), an arm of the World Trade Organization.

[27]Formerly called the International Office of Epizootics (OIE).

BOX 1.9
Non-Biological Factors for Select Agent Designation—Smallpox and Polio

Smallpox was declared eradicated by the World Health Assembly in 1980, a moment that is recognized as one of the most important achievements of humankind. Routine smallpox vaccination ceased in the United States in 1980, and earlier in some countries. Almost half the world's population is currently immunologically naïve to the disease. At the same time, the rise of diseases such as HIV/AIDS that weaken the immune system, as well as the prevalence of atopic dermatitis (the vaccine is contraindicated for individuals with atopic dermatitis), would make resumption of routine vaccination difficult. There are today no licensed therapeutics for the treatment of smallpox, and currently licensed vaccines, while effective, are contraindicated for immunocompromised individuals. Historical anecdotes, while not confirmed, suggest that contaminated materials could be used to spread smallpox in target populations. Both the United States and the Soviet Union have engaged in research aimed at weaponizing smallpox (NRC 2009b). While variola's virulence makes it a threat to public health, these other non-biological factors make smallpox an obvious candidate for use as a bioweapon. Thus, both biological and non-biological factors contribute to this once endemic pathogen being designated a Select Agent. In contrast, polio is not a Select Agent despite its ability to cause crippling disease and death. Polio virus is endemic in some countries, and is therefore difficult to restrict. Moreover, the availability of effective vaccines reduce the threat posed by this virus. However, if vaccination were to cease, this virus could be viewed as a potential bioweapon and designated as a Select Agent. Clearly, the non-biological context can never be predicted from the viral sequence.

rently feasible to predict the biological characteristics from sequence, it is not even theoretically possible to predict the non-biological considerations from sequence. Ultimately, designation of a microorganism as a "Select Agent" is a judgment call and a policy decision. "Select Agent-ness" is not a strictly biological property.

Biological Criteria

With the multiple non-biological factors that contribute to Select Agent designation, it is clear that "Select Agent" is a regulatory term, rather than a biological one. However, the central criteria for a Select Agent are biological attributes. At the time when the Select Agent list was devised by DHHS, some 60 or so full bacterial genomes were known and annotated as well as a score of viruses.[28] Hundreds of bacterial genome sequences are now available for com-

[28]The first complete sequence of a bacterial genome was published in 1995.

parative genomics. Genomic information was not employed in the initial definition of Select Agents. However, from the outset it was asked, to what degree can genomic sequences be used to detect Select Agents now and, for the future, to what degree can genomic sequences be used to predict potentially dangerous sequences in naturally occurring, genetically manipulated or synthetically derived microorganisms? These questions are the focus of this committee.

In examining the sequences from several hundred bacterial genomes, it is impressive just how much diversity is seen—even in different genomes from the same bacterial species. If one considers the general biological criteria used to designate a microorganism as a Select Agent and then asks, to what degree could these attributes be deduced from sequence, the answer is quite clear. Prediction of pathogenicity, transmissibility, ease of preparation, and ease of dissemination is not possible now or in the foreseeable future (see Table 1.1). In part, this is because we lack the basic biological information, and in part it is because our current predictive algorithms are not sufficiently robust.

Infection of a susceptible host by a Select Agent may lead to morbidity and mortality by many different mechanisms, be it "hemorrhagic fever" due to some viral infections or neurological disease following exposure to a bacterial toxin. With the exception of some toxins, the genetic basis for the disease or death that may follow infection with or intoxication by a Select Agent is, however, not well defined. Pathogenicity seen in a susceptible host is the result of a complex interaction between a pathogen and a host defense system, as well as an environmental context (e.g. age, sex, nutrition, health, immune status, and others). **As will be discussed subsequently, pathogenicity of an organism may be the result of a specific sequence and gene, or more frequently the result of interactions between several genes, various sequences, structural characteristics, and host characteristics. There are too many variables involved on the host side alone to be able to accurately predict whether any given nucleic acid change in the pathogen will involve greater or lesser pathogenicity.** The complexity of these systems argues against a simple gene-sequence basis for "predictive oversight of Select Agents" without substantial new information.

As will be discussed in Chapter 2, there is no current way in which a complex biological factor such as pathogenicity can be predicted from genome sequence. Predicting the function of an individual protein, or single microorganism is daunting. Moreover, the nature of infectious disease is such that accurate prediction of microbial pathogenicity is not possible without information concerning the host and the environmental context.

2

Challenges of Predicting Pathogenicity from Sequence

INTRODUCTION

Our committee was asked to "identify the scientific advances that would be necessary to permit serious consideration of developing and implementing an oversight system for Select Agents that is based on predicted features and properties encoded by nucleic acids rather than a relatively static list of specific agents and taxonomic definitions." It is true that the microorganisms and toxins that currently make up the Select Agent list are defined by taxonomy and by their perceived importance to public health and security. They are (or are products of) pathogens, that is, microorganisms capable of causing disease. Most Select Agents are not typical of the common pathogenic microorganisms seen in human or animal medicine, or in agricultural practice. But Select Agents and more commonly encountered pathogenic microorganisms do share a number of biological properties. **It is essential to understand that pathogenic microorganisms are not defined by "taxonomy"; it is very common for a given microbial species to have pathogenic and non-pathogenic members.** *Escherichia coli* is found in the colon of virtually all humans and animals and is part of their indigenous flora. They are typically harmless. However, a genetically defined, and more recently sequence-defined, subgroup of *E. coli* is the most common cause of urinary tract infection in humans and dogs. From a taxonomic standpoint the microorganisms are unequivocally called *Escherichia coli*; from a genetic and sequence homology standpoint they are distinct categories of *E. coli*. Similarly, the taxonomic genus *Yersinia* includes *Y. pestis*, the causative agent of plague, and other *Yersinia* species that are certainly enteric pathogens but are not Select Agents, and other *Yersinia* species that are not known to be pathogens. The pathogens and the non-pathogens are not distinguished by taxonomy but can now be distinguished reasonably well with genetic and molecular analysis.

Is there a potential for developing and implementing an oversight system for Select Agents that is based on features and properties encoded by nucleic acids? The general answer to this question is yes. The committee believes, however, that the entire concept of "predictive" oversight is flawed in that (1) the current Select Agent list has a non-biological as well as a biological basis for existence and (2) functional "prediction" alone cannot provide a level of certainty sufficient to designate a microorganism as a Select Agent, whose possession is legally restricted. Nevertheless, oversight of novel pathogens, whether natural or synthetic, is clearly seen by policy makers and legal experts to be a necessary component of a comprehensive biosecurity strategy. We propose to discuss here only the biological factors relevant to establishing a sequence-based oversight system that is focused on identifying genes and gene products that are likely to be involved in survival and persistence of a microorganism and its interaction with a host. That would include genes of Select Agents, but would also include a far greater number of genes that are associated with pathogenicity (the ability to cause disease) and virulence (the degree of pathogenicity encoded by a given gene or group of genes). Understanding the basis of such an oversight system requires some understanding of the biology of pathogenicity and of the current limitations of genomic analysis.

THE ART OF SEQUENCE-BASED PREDICTION

It is clear that we are immersed in an age of genomics. As of December 27, 2009, the Web site Genomesonline.org (Genomesonline 2009) reported that 3,606 bacterial genomes were being sequenced and that complete DNA sequences of at least 712 distinct bacterial strains were in the public domain. The completed sequences include all the bacterial Select Agents and most common pathogens of humans, animals, and plants. Entrez Genomes contained 3,498 reference sequences for 2,374 viral genomes, including all of the Select Agents and common plant and animal viral pathogens.

The genomes of prokaryotes possess specific and relatively well-understood promoter sequences (signals), such as transcription factor binding sites, that are relatively easy to identify. The gene sequences that code for a protein occur as one contiguous open reading frame (ORF), which is typically many hundreds or thousands of base pairs long. The nucleotide compositions and frequency of use of stop codons (the punctuation between genes) are well known. Furthermore, protein-coding DNA has periodicities of occurrence and other statistical properties. Therefore, recognizing genes in prokaryotic systems is relatively straightforward, and there are well-designed algorithms to do it with high levels of accuracy.

However, identifying a gene and understanding its function are altogether different matters. At least one-fourth of genes that are identified in bacterial genomes, whether large or small, whether from pathogen or non-pathogen, are "hypothetical" or of unknown function. Considering the long history of biochemical and genetic examination of microorganisms, it is daunting that so much of the "equipment" of microorganisms is still unknown. The hypothetical genes are in two categories: ones that are found in a variety of organisms, and ones that are peculiar to particular lineages. For many genome sequences, the only annotation that will be available for the foreseeable future will be based on computational predictions and comparisons with known functional elements in related microorganisms. Despite the hypothetical genes, it has been pointed out that much of the whole-genome sequence of many microorganisms, when viewed against a backdrop of 100 years or more of biochemistry and microbiology, is not at all a total surprise in that much that has been found was fully expected (Doolittle 1981). Moreover, the full genome itself is a tremendously valuable asset because for a given organism, it provides all the details: The entire "parts list" is on the table—even if we do not know where some of them go or what they do. Seeing an assemblage of parts should not be mistaken for understanding how the parts function. From the standpoint of the present report, for important pathogens, comparisons between strains can pinpoint differences between the virulent and the avirulent microorganisms, and comparisons between species can be informative about host or tissue specificity. Comparisons have become even more useful as we have factored in the complete genomes of the human and other animals that serve as microbial hosts. At the genetic level, genome comparisons begin to reveal the fundamental divergences of microbial life and their evolutionary origins. We have also begun to understand how pathogens got to where they are and, we know to some extent what to look for if we are trying to design a pathogenic microorganism.

The objective of a "predictive oversight" system would be to forecast with a high degree of certainty the pathogenic potential of sequences of

- **Single or small numbers of genes related to Select Agent toxins.**
- **Genomes or genomic regions that are closely related to Select Agent pathogens.**
- **Genomes or genomic regions of newly identified natural pathogens.**
- **Novel genomic sequences that are designed and assembled by synthetic biology.**

Sequence prediction in biology is a hierarchy of increasing difficulty that reflects the complexity of the particular system under analysis. The simplest of such predictions would probably be that of a protein, such as a toxin.[1] Next

[1]This is by no means easy. For instance, Yoshida et al. have shown that three amino acid changes can turn the *E. coli* major chaperone GroEL, into an insect toxin. That co-option of function presents a major problem for predictive systems, even at this level. Yoshida, N., K. Oeda, et al. (2001). "Protein function: Chaperonin turned insect toxin." *Nature* **411**(6833): 44.

in order of predictive difficulty would be a genetic pathway (a group of co-regulated multiple proteins interacting in concert). The third most problematic set of sequences to evaluate as a means of forecasting function would be those of whole organisms alone in a controlled environment (with multiple pathways interacting in concert). The final and most difficult predictive situation would be one in which two or more organisms interact in their natural environment.[2] It is this last level of complexity that gives rise to the key biological attributes of pathogenicity and transmissibility, factors that contribute to the criteria that form the basis of inclusion of an organism on the Select Agent list.

Predicting pathogenicity or transmissibility of a microorganism requires a detailed understanding of multiple attributes of both the pathogen and its host. It is a prediction problem of the greatest complexity. Using a single genomic sequence to predict the potential consequences of the interaction of a microorganism, or a microbial virulence determinant, with a host clearly is not within the bounds of contemporary biology. Current sequence prognostication methods are at best at the level of foretelling the function of an individual protein on the basis of its deduced amino acid sequence. Even with the availability of a high-resolution protein structure, projecting the activity of closely related molecules accurately is not straightforward. There is as yet little work that even attempts to make predictions at the next level, that of genetic or biochemical pathways.

Predicting Biological Function from Sequence

The integration of experimental and computational information suggests that the human genome encodes about 20,000 protein-coding genes and an unknown number of functional RNA molecules; the *Bacillus anthracis* (anthrax) genome encodes about 6,000 proteins; a large virus, such as the smallpox virus, encodes about 200 proteins; and small positive-strand or negative-strand viruses, such as coronaviruses and influenza viruses, encode 10-30 proteins. As noted above, although these expressed RNAs and proteins can be identified using computational approaches with relative certainty, assignment of function is problematic. Biological experiments are still needed to confirm computational predictions.

The dominant method of function "prediction" uses sequence homology software. The underlying principle of such an approach is that proteins are reused or modified for applications in similar functional systems in different species far more often than entirely new ones are introduced. Most proteins generally fall into a relatively small number of homologous protein families of related structure and usually of at least somewhat related function. For

[2]Consider the enormous number of gene sequences at play and which must be choreographed as a microorganism leaves the salivary gland of a biting insect and is injected into human tissues.

example, the Pfam protein family database contains about 10,000 protein families that account for about 75 percent of all known proteins. Two proteins that diverged through evolution from a common ancestral sequence ("homologous" sequences) tend to have structural and functional characteristics in common. The sequence that governs the mechanism of action of a particular protein evolves slowly, whereas the sequence that affects how a protein interacts with a binding partner, such as a cellular receptor, evolves rapidly.[3] If the function of one protein is known, some aspects of the functional annotation can be inferred for other homologous proteins. Computer programs for sequence-database homology search (such as BLAST, HMMER, and FASTA) are widely used to discern whether a newly annotated protein or RNA sequence is homologous to an already known sequence or sequence family.

Homology offers only a "low-resolution" prediction of function. **Sequence-homology analysis can often determine what a protein is likely to do (*such as*, protein kinase, metalloprotease, or oxidoreductase) but generally will not reveal the biochemical pathway to which its proteins partner(s) belong or the particular residue(s) that will be the target or substrate for it.** There are less well-developed computational prediction methods that may occasionally offer clues to help to answer the more detailed questions but, generally, such queries must be addressed directly with controlled laboratory experiments. For example, if a novel influenza-like genome were obtained, sequence analysis would certainly and immediately recognize the homologous parts: the genes that encode hemagglutinin (HA) (Pfam protein family database code PF00509), neuraminidase (NA) (PF00064), nucleoprotein (PF00506), the matrix proteins M1 (PF00598) and M2 (PF00599), the proteins NS1 (PF00600) and NS2 (PF00601), and the RNA-dependent RNA polymerase components PA (PF00603), PB1 (PF00602), and PB2 (PF00604). These families have tens of thousands of examples of sequences; on the basis of the known diversity, the statistical models used in sequence-homology analyses are often capable of recognizing sequences that are separated by hundreds of millions or even billions of years of evolution. The computational approaches would identify the general kinds of "parts" in a genome and would be able to determine whether an expected part were present, missing, or unexpectedly quite different from a currently known virus sequence component. We can recognize some molecular signatures that may be essential for maintaining effective pathogen-host inter-

[3]For example, a particular enzyme cleaves DNA and recognizes a specific sequence defined cleavage site; the enzyme structure that allows it to cleave DNA may evolve slowly (related enzymes also cut DNA), whereas the portion of the enzyme that recognizes the specific DNA sequence cleavage site might evolve rapidly (related enzymes cut different DNA sequences).

action,[4] replication efficiency and pathogenesis outcomes in natural, but not necessarily, alternative hosts. The identification of genera or species of pathogenic bacteria or viruses is likewise easily accomplished with DNA homology approaches. Those methods are used every day in clinical laboratories all over the world.

What could not be readily foretold from the sequence-homology analyses described above is whether the influenza-like virus is highly pathogenic for humans and other mammals, or whether a particular vaccine will protect against it. Those traits depend on a small number of genetic changes that evolve rapidly in ways that are not well understood; even subtle changes may have a profound biological effect. Those features change so rapidly, they do not correlate well with evolutionary history. Thus, sequence-homology analysis is less informative for such viral characteristics than for simply identifying genome parts components of an influenza or related virus. For example, there is a strong correlation between high pathogenicity and trypsin-independent cleavage of the influenza virus hemagglutinin. This is not a perfect means of prediction, however, inasmuch as the 1918 influenza virus, which was associated with 50 million deaths worldwide, has a cleavage site that appears from sequence analysis to be associated with low pathogenic potential (Box 2.1). **Such poor predictive power from sequence analysis is likely to be common for many *if not most*, microbial virulence determinants because virulence is typically mutifactorial and is affected by details of molecular interactions between a microorganism and a specific target in a specific host.** (See also Appendix G)

Protein Structure Prediction

Another possible route to prediction of function from sequence is to predict the folded 3D structure of a protein from its sequence and then use features of that structure (which evolve much more slowly than sequence) to infer catalysis, binding partners, or other functional properties. Function prediction based on structure has been one of the "Grand Challenge" problems in science for the last 50 years, since Anfinsen showed that the information to determine protein 3D structure is encoded in the linear amino-acid sequence (Haber and Anfinsen 1962).

Pure *de novo* structure prediction was essentially impossible until recently and is only occasionally successful even now. Progress has come mostly from the growing database of experimentally determined structures (the Protein

[4]Methods such as CorrMut and CRASP identify functional domains within proteins that are co-evolving in response to one or more unknown selective pressures (Afonnikov, D. A. and N. A. Kolchanov (2004). "CRASP: a program for analysis of coordinated substitutions in multiple alignments of protein sequences." *Nucl. Acids Res.* **32**(suppl_2): W64-68, Fleishman, S. J., O. Yifrach, et al. (2004). "An evolutionarily conserved network of amino acids mediates gating in voltage-dependent potassium channels." *Journal of Molecular Biology* **340**(2): 307-318.

BOX 2.1
Influenza—Hemagglutinin Cleavage

One of the most important sequence features of influenza virus A pathogenesis is a protease cleavage site in hemagglutinin (HA). Cleavage is required for HA to catalyze membrane fusion, a necessary step for viral infectivity. In viruses of low pathogenicity, this essential cleavage step tends to be catalyzed by a host-encoded protease (trypsin, in the human respiratory tract), so viral infectivity is limited by tissue distribution of the host protease. Conversely, an essential feature of highly pathogenic influenza viruses is the presence of mutations in the gene for HA that leads to trypsin-independent cleavage of the protein; such mutations enable the virus to infect a broader range of tissues. The HA protease cleavage site is a sequence of only about seven to 10 residues, and the sites that contain small insertions of a few basic residues (lysine and arginine) tend to be associated with trypsin independence. With such changes, HA becomes cleavable by widely distributed subtilisin-like proteases that have a consensus recognition sequence. Yet, the correlation of high pathogenicity and trypsin independent cleavage of HA is not perfect; the 1918 influenza virus, which was associated with 50 million deaths worldwide, has a cleavage site that appears from sequence analysis to be of the low pathogenic.

Data Bank contains over 60,000; http://www.rcsb.org/pdb), which enable the modeling of new sequences on the basis of homology to known, related structures. Recent achievements have been impressive, as demonstrated by the Critical Assessment of protein Structure Prediction (CASP) competitions (Moult 2005). Two rather distinct current approaches achieve good predictions fairly often: One is inspired by protein evolution and can recognize and piece together distant sequence relationships (Zhang, Wang et al.), and the other is inspired by protein folding and uses a combination of physics and empirical data to construct a model (Raman, Vernon et al. 2009). Homology models are usually approximately correct (Keedy, Williams et al. 2009), and even *de novo* predictions sometimes succeed (Box 2.2).

Further prediction of binding or catalytic sites from successfully modeled 3D structures (López, Ezkurdia et al. 2009) or prediction of protein/protein binding modes by docking known components, as in the CAPRI (Critical Assessment of PRedicted Interaction) competition (Janin 2005), are also still successful only sometimes and only partially. Approximately correct homology models can enhance the power of purely sequence-based comparisons considerably, especially when they show that known functional residues are brought together into the right 3D relationships. Often, however, the critical biological details hinge on structural details that confer a difference in specificity or in regulation and that are exactly the most difficult places to achieve accurate prediction.

BOX 2.2
Critical Assessment of Protein Structure Prediction (CASP) Competition

Last year's **Critical Assessment of protein Structure Prediction** (CASP8) results for free modeling (that is, when no clear template is available in the Protein Data Bank) included at least one good model for 5 of the 13 targets (Ben-David, Noivirt-Brik et al. 2009). Some predicted models have been shown to be accurate enough for the demanding application of molecular replacement for solving the crystallographic phase problem (Qian, Raman et al. 2007).

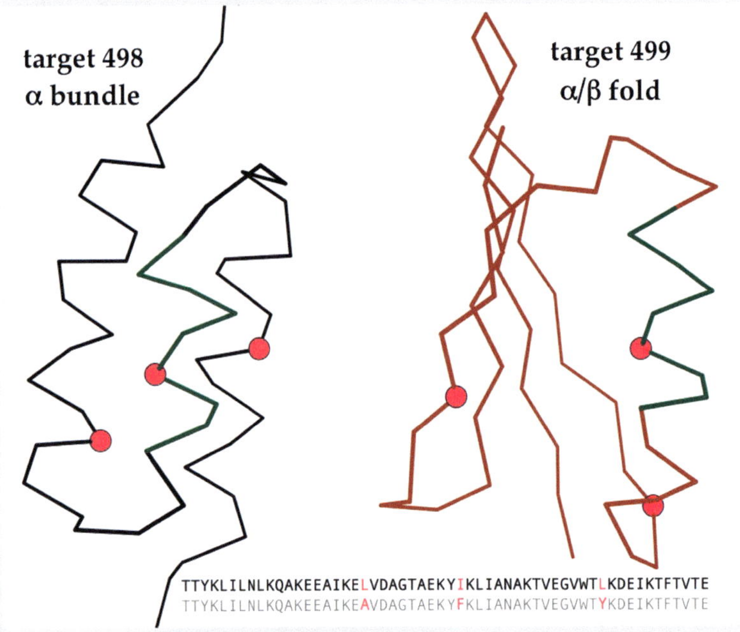

CASP8 included a cautionary example of extreme structure-prediction difficulty in the form of two targets that had only three different residues out of 56 (as shown in the figure); the pair had been designed and selected for maximum sequence match but to fold into entirely different 3D structures (He, Yeh et al. 2005). A purely sequence-based predictive method could not recognize the consequences of this tiny difference, and it seems that the predictor groups who got both targets right knew about the earlier stages of this tour-de-force design (He, Yeh et al. 2005). The example is analogous to the issue of the vaccine strain of a pathogen; in both cases, a very small, "linchpin" change in sequence causes a reversal of the large-scale relevant property: In this pair the protein fold, and for the vaccine strain, its pathogenicity. Pure prediction is therefore chancy at best here, and correct classification depends on expert outside knowledge of such unusual near-neighbor cases.

Overall, this route of prediction based on 3D structure is well worth encouraging for both practical and intellectual benefits, but its utility in a robust system of predicting Select Agents (for legal oversight) is still extremely far off.

Gene Regulation

If an organism's virulence depends on specialized gene products, it must be able to use them when they are needed but not squander its metabolic energy in producing them aimlessly or risk having them detected and prematurely neutralized by host defenses. Consequently, regulating the expression of virulence factors is an additional essential complication of a pathogenic microorganism's life. The host presents an array of conditions strikingly distinct from those

BOX 2.3
Ricin

Ricin is a plant toxin isolated from the seeds of the castor bean plant, *Ricinus communis.* It inhibits protein synthesis in affected cells by modifying the ribosome; this leads to ribotoxic stress and eventually cell death. Ricin represents a family of toxins known as Ribosome Inactivating Proteins (RIPs) that are found throughout the plant and bacterial kingdoms. Despite their very different sequences and sometimes quite different structures, these toxins share three highly conserved amino acids that are responsible for their catalytic activity.

One cautionary tale for the prediction of toxin activity from gene sequence comes from the work of Frankel and Robertus (Frankel, Welsh et al. 1990). They genetically mutated ricin amino acid glutamate-177 and predicted that the protein would be inactivated because this side chain is highly conserved and central to the catalytic activity of the toxin. Although more conservative mutations were inactive, when the glutamate was mutated down to an alanine residue, the enzyme still retained about 5 percent of the activity of the wild-type sequence—enough to slow growth of yeast cells sensitive to the toxin. Based on the structure of ricin, the researchers predicted that the nearby glutamate residue at position 208 was able to move and substitute in the reaction. They produced an inactivated ricin A chain only after mutating both glutamates 177 and 208; a crystal structure of the Ala-177 mutant showed that the carboxyl of Glu-208 did indeed move into the former position of catalytic Glu-177 (Kim, Misna et al. 1992).

The genes encoding ricin (rtx genes) of the various castor bean cultivars have only one or two nucleotide differences in regions that do not affect protein structure or function. The *R. communis* genome would not be sequenced to determine the virulence of castor beans, but instead it would be assumed that unmodified beans contain active ricin toxin. Several research groups have genetically altered one or more of the conserved catalytic residues to produce inactive ricin A chain expressed recombinantly in *E. coli* in an attempt to produce a vaccine to protect against deliberate ricin poisoning (Munishkin and Wool 1995).

of the outside environment, conditions that are not easily reproduced in the laboratory. In fact, laboratory culture conditions bias our understanding of microbial adaptation to natural environments. *Vibrio cholerae*, for example, is thought to persist without expression of virulence factors in brackish estuaries and other saline aquatic environments, sometimes in association with the chitinous exoskeletons of various marine organisms. Transition from that milieu to the contrasting environment of the human small intestinal lumen is accompanied by substantial genetic regulatory events.

The microbial cell is relatively simple, yet it possesses the means to detect, often simultaneously, changes in temperature, ionic conditions, oxygen concentration, pH, and calcium, iron, and other metal concentrations that might appear to be subtle signals but are essential for the precise mobilization of virulence determinants. Similarly, environmental regulatory signals prepare a microorganism for its transition from an extracellular to an intracellular state. For example, iron is a critical component of many cell metabolic processes; therefore, it is not surprising that animals rely on high-affinity iron-binding and iron-storage proteins to deprive microorganisms access to this nutrient, especially at the mucosal surface. In turn, most pathogens sense iron availability and induce or repress various iron acquisition systems accordingly. Moreover, many microorganisms produce toxins that are regulated by iron in such a way that low iron concentrations trigger toxin biosynthesis. Reversible regulation of the expression of virulence genes by temperature is common to many pathogens. Thus, a microorganism like *E. coli* that may be deposited in feces and live for long periods under conditions of nutrient depletion and low temperature, mobilizes its colonization-specific genes when it is returned to the warm mammalian body. The regulatory machinery used to accomplish that is an important feature of many pathogens, including *Y. pestis* and *B. anthracis*.

The number of well-characterized virulence regulatory systems is rapidly increasing, in part because of the development of rapid methods for screening gene expression on a genome-wide basis (for example, with the use of DNA microarrays). But, relatively little is known about either the specific environmental signals to which these systems respond and or the exact role of the responses in the course of human infection. One common mechanism of bacterial transduction of environmental signals involves two-component regulatory systems that act on gene expression, usually at the transcriptional level. Such systems make use of similar pairs of proteins: one protein of the pair spans the cytoplasmic membrane, contains a transmitter domain, and may act as a sensor of environmental stimuli; the other is a cytoplasmic protein (a "response regulator") with a receiver domain and regulates responsive genes or proteins. Those regulatory systems are common both in pathogens and non-pathogens, so their detection by sequence analysis cannot be used as a reliable predictor of whether a microorganism is pathogenic.

The coordinated control of pathogenicity incorporates a *regulon*; a group of operons or individual genes controlled by a common regulator, usually a protein activator or repressor. A regulon provides a means by which many genes can respond in concert to a particular stimulus. At other times, the same genes may respond to other signals independently. Global regulatory networks are a common feature of microbial virulence and basic microbial physiology, so their sequences, although often essential for a pathogen, are not reliable predictors of virulence. The apparent complexity of virulence regulation in a single microbial pathogen is magnified by the coexistence of multiple interacting ("cross-talking") systems and by regulons within regulons.

Thus, the inherent pathogenicity of a microorganism can be greatly altered through regulation of virulence genes. It is extremely difficult to predict how even a single nucleotide change will affect regulation and thereby alter pathogenesis or the viability of the microorganism. (Additional detailed examples of the important role of regulation in allowing pathogens to respond to the environment of the human host are given in Appendix J.)

THE NATURE OF INFECTIOUS DISEASE AND THE ART OF PREDICTING PATHOGENICITY

The preceding sections have shown that several computational approaches have promise for predicting biological function from sequence. Can they be applied effectively to predict pathogenicity? If not, what is required to develop a predictive method that would suffice? To address this issue, we will first discuss the nature of infectious disease.

Infectious diseases affect all living things, from the smallest amoeba to insects, plants, and the largest mammals. The co-existence and co-evolution of microorganisms with their hosts is a dynamic equilibrium ranging from one extreme of mutualism in which both partners benefit from the interaction (for example, bacterial production of organic nitrogen for plants or of vitamin K for the human), to a relationship of commensalism in which one organism benefits but the other is unaffected, to another extreme of parasitism that benefits one partner to the detriment of the other.

Microorganisms are constant companions of plants and animals. Humans carry a vast indigenous microbial flora from shortly after birth until death, and the role of this human microbiome in human health and disease is the subject of considerable interest and recent investigation. Although it is biologically correct to say that most microorganisms that inhabit this planet are harmless to humans or may even benefit humankind, it is also true that humans have a prejudicial view of microorganisms and direct their focus to microorganisms as agents of disease. The biological reality is that most microbial infections are relatively benign and that symptoms of disease are sometimes the result of the human immune system's response to infection rather than the product of the infecting microorganism (Box 2.4).

BOX 2.4
Categories of Microorganisms

For humans, and to a lesser extent other animal and plant hosts, interactions with microorganisms are complex and are governed by the health status of the host and by the environment. On the basis of differences in such interactions, microorganisms can be categorized as follows:

- ***Transients.*** Microorganisms encountered daily in food or in the environment. Generally, these microorganisms are just "passing through" and are of no consequence.
- ***Commensals*** (derived from the Latin meaning to eat from the same table). The vast community of microorganisms (which make up the microbiome) that normally inhabits the host. These are generally harmless. In fact, there are about 10 bacterial cells for every human cell in the body. Recent studies suggest that the human microbiome may be as distinct as a fingerprint, although the general microbial components are similar among individuals.
- ***Pathogens*** (derived from Greek words meaning the birth of pain). Pathogens may or may not be members of the commensal flora, but they are notable because they regularly can cause disease in apparently healthy individuals.
- ***Opportunistic pathogens.*** Microorganisms that *only* cause disease in human hosts that are in some way compromised in their normal immune defense mechanisms. Host defense may be compromised as a result of: pregnancy; age (young and old); an underlying disease (such as, diabetes or malignancy); therapies for malignancy or organ transplantation; or traumatic injury or burns. The nature of the immunodeficiency will dictate the particular types of microbial infections.
- ***Accidental pathogens.*** Microorganisms that are encountered in accidental contact with other animals, insects, or the environment. Many of these microorganisms are among the most deadly for humans and are disproportionately represented on the Select Agent list of potential bioterrorism agents. Generally speaking, these microorganisms are distinguished from human-specific pathogens in not being directly or readily transmissible from one human host to another. A human host is in essence a biological "dead end" or incidental host. Nonetheless, many of the general classes of virulence factors found in accidental pathogens are also found in human-adapted pathogens.

Recognition of an organism as a pathogen is not always simple, because the interplay between genetic expression of the microorganism and the host is ever-changing. However, in the biological sense, pathogens possess the ability to cross anatomic barriers or breach other host defenses that limit the survival or replication of other microorganisms. The more complicated and relevant question, however, is why some microorganisms are pathogenic to humans or other animals, yet other closely related microorganisms are not. That question can be approached from the point of view of the host or the microorganism:

- What specific defense mechanisms of the host allow it to effectively suppress infection (entry, attachment, invasion, and replication) by certain microbes and not others?
- What are the differences between microbial agents that cause disease and those that do not?

Pathogenicity depends on the biological context, so pathogenic traits (virulence factors) need to be defined in terms of their potential to be associated with infection and disease in a particular host. For example, the human gastrointestinal epithelium is exquisitely susceptible to cholera toxin when delivered during infection by *Vibrio cholerae*. Although the gut epithelium of most animals is also susceptible to the action of the toxin, the disease cholera is seen only in humans: This finding suggests that additional host-specific virulence factors are involved. The vulnerabilities of the mammalian immune defense mechanisms to factors used by microorganisms to exploit or overcome a specific host defense strategy are of critical importance in defining what is and what is not a microbial virulence factor. **To restate the point simply, the nature of microbial virulence factors cannot be understood unless the factors are evaluated in the context of the biology of the host.**

Most genes encoded by a pathogen are necessary for its replication and for "general housekeeping"[5] and genes are in general shared by most microorganisms. Virulence genes can be considered to be specialized genes essential for survival of a particular organism in a particular environment, usually on or in a host. Even by that limited operational definition, most pathogenic and non-pathogenic bacteria and large viruses that inhabit animals and plants have large numbers of virulence genes. For example, the variola virus (smallpox virus) virulence gene family consists of 64 genes (of a genomic total of 193); however, the presence of a large number of virulence genes does not *a priori* guarantee pathogenicity. Camelpox and taterapox viruses are the closest orthopoxvirus relatives of variola virus and have similar number of virulence genes, but neither virus causes human disease. Cowpox virus has a larger genome than does variola virus and encodes a greater number of virulence genes, but it causes only a localized lesion in humans and was used by Jenner in the late 1700s to vaccinate against variola virus. Those observations illustrate that variola virus

[5]In addition to the immune defense system, pathogenic microorganisms face both unique and common obstacles during infection of human, animal, and plant hosts. The genes important during the natural life cycle of the pathogen include those for: (a) entering a host; (b) finding a unique niche within the host; (c) interacting intimately with host tissue or cellular factors; (d) avoiding host defense mechanisms; (e) replicating or persisting within a host; (f) synthesizing toxins and other virulence factors; and (g) exiting and disseminating from the host.

pathogenicity is a complex trait, possibly due in part to subtle activity differences between the virulence genes of viruses that are pathogenic for humans and the genes of viruses that are not pathogenic for humans. Some host specific pathogenicity can result even from the loss of function (see Appendix J). In the case of bacteria, the picture is a little clearer inasmuch as bacterial pathogens usually have virulence genes that are not present in their non-pathogenic relatives, and this distribution suggests that bacteria evolve to become pathogens by acquiring virulence determinants. For example, *Salmonella* and *E. coli* evolved from a common ancestor, but *Salmonella* acquired genes distinct from those of *E. coli* that permitted it to cross the mucosal barrier of the gastrointestinal tract. In contrast, *Yersinia pestis* adapted to life in a flea and in rodents and has retained most of the genes contributed by its most recent ancestor *Y. pseudotuberculosis*, which is an enteric pathogen. Some of the *Y. pseudotuberculosis* genes have been inactivated; they are not necessary for *Y. pestis* survival or they interfered with its new "lifestyle." In parallel, microbe–host interactions play a critical role in regulating disease severity and distribution in human, animal, and plant populations. It is now generally accepted that pathogenic microorganisms have shaped the genetic population structure of humans and vice versa (see Appendix H).

What Is the Origin of Bacterial Pathogenicity? What Makes a Pathogen?

Microbial evolution continues to challenge the state of human health in part because of the size of the microbial universe; macroscale changes in human, animal, and plant interactions; and the dynamic nature of genomic alterations that result from active gene flow between microorganisms. The net effect of evolution on an organism is a balance between selective pressure in the environment of that organism and the generation of relevant changes in the microorganism through mutation, gene duplication, and horizontal gene transfer.

Most medically important microorganisms have pathogenic and non-pathogenic strains, and virulence factors were well known and characterized long before the advent of the genome projects. However, what we have learned about pathogenicity from the examination of microbial chromosomes has been surprising and useful. **One of the most important findings both from the perspective of understanding the biology of pathogenicity and from the standpoint this committee's task, is that pathogenic bacteria often contain clusters of genes, called pathogenicity islands (PAIs), that are not present in related non-pathogenic bacteria.** Those acquired genes have several features (for example, the G+C content of their DNA and other molecular signatures) that indicate that they were once associated with mobile genetic elements, such as bacterial plasmids or bacteriophage (viruses that infect bacteria). Thus, it is clear that although *Salmonella* and *E. coli* evolved from a common ancestor, *Salmonella* has acquired genes distinct from *E. coli* that permit it to invade gut epithelial cells. As noted, *E. coli* strains vary greatly in that some commonly cause disease whereas the majority of *E. coli* strains are universally restricted to the commensal flora and never cause harm in an immunocompetent host. However, *E. coli* that "routinely" cause infections, such as urinary tract infections, and the notorious *E. coli* O157, have genes, indeed large clusters of genes, that are associated with their pathogenicity and are not found in non-pathogens. We now understand that uropathogenic, enterohemorrhagic, and extraintestinal types of *E. coli* all display mosaic genome structure, with hundreds of gene islands distinct to each type, that makes up as much as 40 percent of the overall gene content in each of these strains. Pathotypes are as distinct from one other as each is distinct from a nonpathogenic laboratory strain of *E. coli*. *E. coli* has an "open" pan-genome: with every new genome sequence, a new set of about 300 unique genes is discovered; this suggests continuing evolution of this species by gene acquisition. The O157 case is a good example of how pathogenicity requires biological context; this microorganism is asymptomatically carried by animals, but it can cause diseases ranging from simple, barely noticeable diarrhea to devastating colitis and death in humans.

In contrast, *B. anthracis* and other pathogens that have restricted environmental habitats display a "closed" pan-genome and a much greater fraction of shared genes. Nevertheless, *B. anthracis* contains two large plasmids, one of which has a 44.5-kb pathogenicity island that contains genes for (among other things) the toxin that can be lethal to humans. It also contains a sequence signature that suggests a history of gene shuffling and exchange. It is therefore not too surprising to learn that the sequence of the main chromosome of *B. anthracis* is remarkably similar to those of *B. cereus* and *B. thuringiensis*, which also inhabit the soil. *B. cereus* strains carrying part of the pathogenicity apparatus of *B. anthracis* are occasionally encountered and present a problem for those attempting to draw a line separating Select Agents from non-Select Agents. Although it may be tempting to think of microorganisms in a clear-cut pathogen and non-pathogen manner, the biological reality is much more complex. Far from black and white, microbial genomes are better described as a multicolor patchwork of sequences, see Figure 2.1 (Read, Peterson et al. 2003). It is also instructive to see that *B. thuringiensis* produces a toxin fatal to lepidopteran caterpillars; thus, there is a biological theme at play that involves microorganisms, their hosts, and the common mechanisms that microorganisms use to ensure their survival. The distinction of *B. anthracis* from related soil microorganisms, such as *B. thuringiensis* and *B. cereus*, is not always clear in a sequence homology context, whereas the sequence information makes good sense in a biological context. For a more in-depth discussion of pathogenic mechanisms and virulence genes with examples of Select Agents, see Appendix I.

It is perhaps difficult for non-specialists to appreciate that the smallest free-living organisms on the planet engage in a kind of primitive "sex life" that

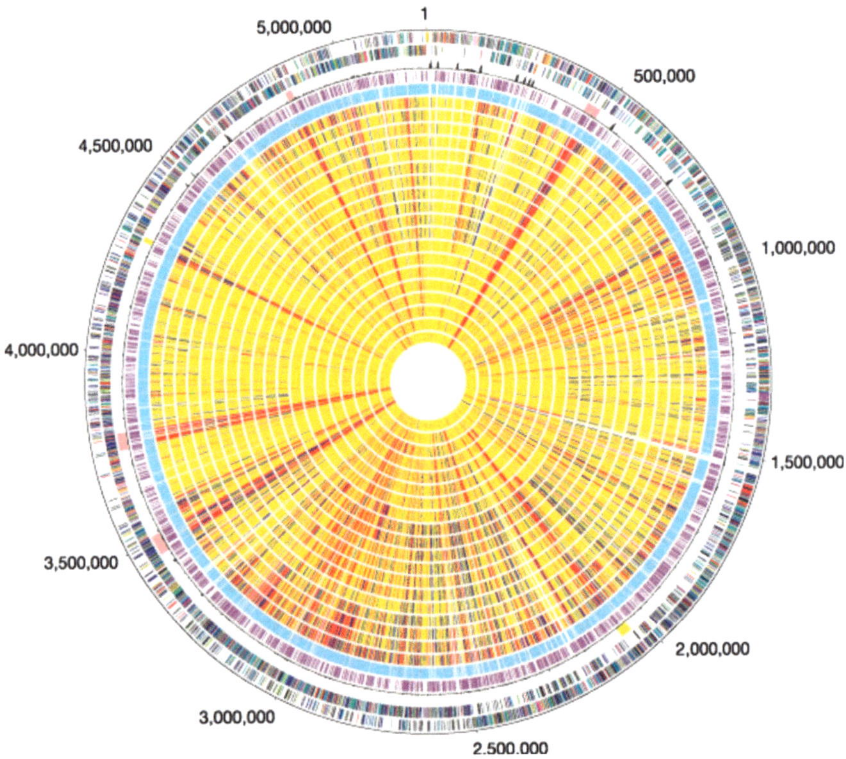

FIGURE 2.1 The genome sequence of *Bacillus anthracis* Ames and comparison to closely related bacteria. Outer circle, predicted coding regions on the plus strand colour-coded by role categories. Circle 2, predicted coding regions on the minus strand colour-coded by role categories. Circle 3, atypical nucleotide composition curve. Salmon colour, phage regions; yellow, other unique regions located around positions 2.0 and 4.3 Mb. Circle 4, genes not represented on the array. Circle 5, genes present on the array. Genes were classified into three groups: genes present in the query strain (shown yellow), genes absent in the query strain (red), and diverged genes (blue). Missing data are in grey. *B. cereus* group strains are displayed following the phylogeny of Fig. 2.1 (circle number, strain number): 6, *B.c.* 874; 7, *B.c.* 535; 8, *B.c.* 612; 9, *B.w.* 1143; 10, *B.t.* 248; 11, *B.t.* 442; 12, *B.c.* 14579; 13, *B.t.* 775; 14, *B.c.* 259; 15, *B.t.* 1031; 16, *B.t.* 251; 17, *B.c.* 607; 18, *B.c.* ATCC 10987; 19, *B.c.* 812; 20, *B.c.* 819; 21, *B.c.* 831; 22, *B.t.* 840; 23, *B.c.* 1123; 24, *B.c.* 816. Here we use *B.c*, *B.t.* and *B.w* to indicate *B. cereus*, *B. thuringiensis* and *Bacillus weihenstephanensis*, respectively (Read, Peterson, et al. 2003).

involves a rather promiscuous exchange of genetic information called lateral (or horizontal) gene transfer (see Box 2.5). The public is already aware that the emergence of antibiotic resistance in microorganisms is a serious health issue, and the mass media have reported that this emergence of resistance is effected by gene exchange among bacteria in which resistance is "donated" to previously susceptible microorganisms. The same is true of pathogenicity, although the transfer of genes associated with pathogenicity is not a single event. Rather, the genomic analysis of pathogenic microorganisms reveals that there have been multiple such events over a long period and that many other genetic selection events have occurred to finely hone only a relatively few microorganisms to emerge as "pathogens."

A discussion of the current understanding of the emergence of pathogens based on the study of their genomes is beyond the scope of this report, although it can be readily found elsewhere by an interested reader. For present purposes,

BOX 2.5
Acquisition of Virulence Factors—Horizontal Gene Transfer

Horizontal (or lateral) gene transfer is any transfer, exchange or acquisition of genetic material that differs from the normal mode of transmission from parents to offspring (vertical transmission). Lateral gene transfer can occur by several mechanisms:

- Transformation: acquisition of DNA molecules.
- Transduction: gene transfer mediated by a virus.
- Conjugation: DNA transfer by direct cell-to-cell contact.

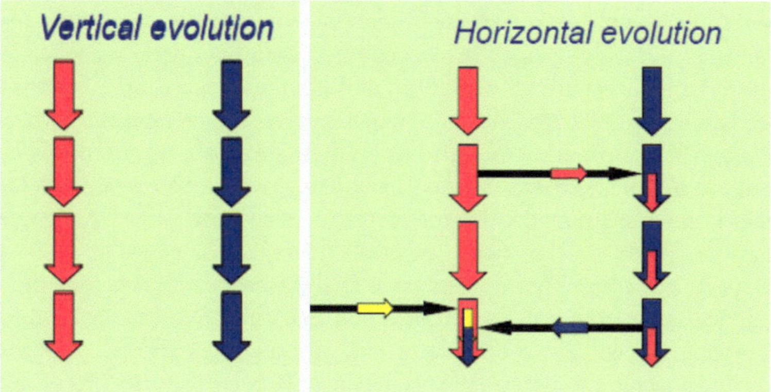

These mechanisms cause the transfer or acquisition of genes among members of the same species or related species or between members of different taxa.

it is sufficient to state that it is sometimes possible to recognize genes that arose by lateral gene transfer by simply examining genome sequences. The evolution of pathogenicity is a continuing process that has in some cases been aided by unintentional changes caused by human activity. Moreover, what we have viewed as progress in medicine and agriculture has provided new opportunities for microorganisms to adapt to humans and other organisms and to cause substantial morbidity and mortality (for example, overuse of antimicrobials leads to the selection and transfer of resistance).

A fundamental evolutionary push to bacterial pathogenicity results from gene acquisition. That is not simply a mechanism whereby microorganisms become pathogenic but a general strategy for specialization and success in some environmental niches that are highly competitive. What is the origin of the inherited genes that go on to be important for pathogenicity? The answer is not clear; microorganisms do not leave a fossil trail. However, what is important to our task is that to some extent their history can be read in their sequences. Bacterial adaption of the tactic of horizontal gene transfer to maximize diversity and to increase the opportunity for continuing evolution is most likely a reflection of their haploid state; that is, bacteria and viruses have only a single chromosome, so any genetic change, either beneficial or detrimental, is immediately apparent. Microorganisms have a difficult balancing act. They have the need to conserve fundamental characteristics while being able to try new combinations of genes. The sharing of genes among seemingly disparate microorganisms that occupy the same niche provides these microorganisms with an endless number of combinations of genes for evolutionary experimentation. If we trace the kinds of genes that appear to be peculiar to pathogens, we see that similar kinds of genes encode toxins or invasins found in organisms that infect plants, people and organisms in between. To some extent, we find that virulence genes evolved to prevent predation. For example, there are similarities between a microorganism being devoured by a free-living amoeba, which may have happened early in evolution, and being devoured by a macrophage in the alveolus of a mammalian lung, which evolved much later. In both cases, some organisms' formation of capsules on their surfaces prevented this event; indeed, the complete genome sequences of many pathogens reveal that clusters of genes in the pneumocococcus, the meningococcus, and some SA are needed for capsule synthesis. That does not mean that capsular genes in all bacteria are suspect, but contemporary sequence analysis can tell us whether the capsular genes come from a pathogen or a non-pathogen. We can also trace how microorganisms evolved to be able to live within phagocytic cells. *Legionella pneumophila* prefers to parasitize and live in amoebae in nature but, under the right circumstances, can survive and replicate in the phagocytic cells of a human lung. *Mycobacterium tuberculosis* parasitizes human alveolar macrophages preferentially and can live and persist in these cells for a human lifetime. Yet, *Legionella* and mycobacteria utilize recognizably similar general mechanisms for survival in the lung; these mechanisms are revealed in part in the genome sequences of the two organisms. Similarly, we find that many other genetically controlled characteristics can confer virulence on bacteria, including factors that enable the bacteria to attach to and disrupt host cells.

Genomic analysis also led to the discovery that many pathogenic bacteria use similar machinery for injecting proteins into the cytoplasm of the host's cells. That common feature of pathogens of plants and humans follows, in a sense, the same kind of features found in viruses. While the nature of the proteins injected into animal and plant cells (effector proteins) may vary, there is often surprising homology among injection apparati across diverse pathogenic species. Perhaps even more striking was the finding that the sophisticated apparatus, which is akin to an assembled bacterial hypodermic syringe, is often found as part of a pathogenicity island in various bacterial pathogens in phylogenetically distinct taxa. For example, the type III secretion system, has been identified in a variety of fully sequenced bacteria ranging from the obligate intracellular parasite *Chlamydia trachomatis* to the plague bacillus *Y. pestis*. Again, not all type III secretory apparatus genes are associated with pathogenicity, but the combination of a type III secretion system (or similar distinct secretory machinery) and the signature for effector proteins provides an immediate clue that one may be dealing with a pathogen. For more discussion and examples of how microorganisms become pathogens, see Appendix J.

Thus, bacteria have evolved a kind of biological network to exchange their genetic knowledge. That is as important for the evolution of the organisms that we call pathogens as it is for the organisms that fix nitrogen in the soil. The specialization of bacteria to live at the expense of other, more highly evolved organisms is a reflection of shared experiences. Genomic sequencing has revealed that human DNA contains traces of similar kinds of gene exchanges mediated by viruses and other mobile genetic elements. The novelty of this biology is impressive and shows that the road to bacterial, viral and parasitic specialization has ancient roots. **The commonality that is still present in microbial sequences makes the goal of finding unique predictive sequences difficult. But it constitutes the most practical means of identifying pathogenic potential today, and it probably will lead to a much greater predictive capacity in the future.**

The Evolution of Bacterial Host Specificity

If the acquisition of genes by horizontal gene transfer seems to be a driving force for bacterial pathogenicity, one might assume that gene loss plays only a minor role in pathogenicity. However, that is not always the case. The fine tuning of pathogenicity, especially with respect to host specificity, appears often to involve gene loss or gene rearrangement (see section on "Gene Loss" in Apendix J). Gene loss can occur by simple mutation (change or deletion of a single or a few nucleotides), which is often reversible. Irreversible gene

BOX 2.6

- **Bacterial genomes are small and densely packed with genes.**
- **Pathogenic bacteria often contain clusters of genes (pathogenicity islands, PAIs) that are not present in related non-pathogenic bacteria.**
- **Many of these virulence determinants were acquired by horizontal gene transfer.**
- **Acquired genes have several features (such as G+C contents, association with plasmids or phage, and sporadic distributions) that denote their ancestry.**
- **It is sometimes possible to recognize genes that arose by lateral gene transfer by simply examining genome sequences.**
- **The amount of acquired DNA in many bacterial genomes can be substantial.**

loss—the permanent loss of a part of, all of, or many of the genes from the chromosome—is common. For example, as noted above, *Salmonella* spp. have a number of pathogenicity islands that were acquired by horizontal gene transfer. The *Salmonella* group of pathogens is notable for its division into specific types that have a preference for a particular host (animal, bird or reptile). A core of identical genes are found in all salmonellae. However, genome comparisons of host-restricted or host-adapted *Salmonella* spp. (and indeed of other pathogens such as poxviruses) indicate that loss of gene function may be a common evolutionary mechanism through which host adaptation occurs. Gene loss appears to restrict the potential pathogenicity in the host to a more limited or specialized set of interactions, often through specific targeting to particular cell types or organs. For example, long term asymptomatic carriage of *Salmonella*, which are shed periodically and act as a reservoir of infection for susceptible hosts, occurs because the organisms establish a niche in the gall bladder, an organ that is immunologically protected. *S. enterica* serovar Typhimurium causes gastroenteritis in humans (typically food poisoning) but causes a chronic infection of mice where the organism can be carried and shed for the life of the animal. Conversely, *S. enterica* serovar Typhi infects humans exclusively, targets the cells of the reticuloendothelial system, and likewise can be shed asymptomatically for a lifetime (as in the case of "typhoid Mary"). The comparison of *S. enterica* Typhimurium with *S. enterica* Typhi shows that Typhi (and other host-restricted *Salmonella*) harbors a significantly higher proportion of genetic events that are associated with the loss of functional genetic sequences. Modification or loss of some effector proteins can affect whether a *Salmonella* strain will preferentially be successful in infecting and persisting in a human, a chicken, or a rodent.

In addition, genomic analysis has revealed that gene loss suffered by host-adapted pathogens is often reflected by the appearance of pseudogenes, which are sequences of bases in the DNA that clearly resemble the sequences of known genes but differ from them in some crucial respect and have no function. Pseudogenes generally do not regain their function except by extensive recombination and their reversion might not lead to a change in pathogenicity. As the pattern of gene gain through horizontal gene transfer and subsequent gene loss through adaptation has been analyzed, it appears that gene loss may be a mechanism of targeting the invading pathogen preferentially to particular tissues or host cells and avoiding the potential stimulation of non-specific inflammation. For example, in both *Salmonella* and *Yersinia*, gene loss may be involved in the adaptation from a gastrointestinal to a systemic "lifestyle." Similarly, genomic analysis seems to refute the popular belief that *M. tuberculosis* evolved from *M. bovis* through the adaptation of a bovine strain to the human host. In fact, *M. tuberculosis* contains chromosomal segments that have been deleted from *M. bovis*, which raises the converse possibility that humans transmitted tubercle bacilli to animals and those bacilli subsequently evolved into *M. bovis*.

This is not simply interesting basic science and evolutionary biology. As genomic analysis has been broadly applied to pathogenic microorganisms, analysis of irreversible genetic events—such as chromosomal deletions (large sequence polymorphisms), single nucleotide polymorphisms (SNPs), and direct repeat content ("spoligotype") patterns—has permitted us to decipher the phylogeny that has occurred during the evolution of pathogenicity. On one hand, we have begun to uncover and understand the mechanism used by nature to "design" a pathogen; this mechanism becomes a blueprint that may be followed by those who would wish to design therapeutics, and by those with more sinister intent who would wish to design novel pathogens. **The loss of genetic information, as well as the gain of genetic information, can be important, and this adds to the present uncertainty of predicting pathogenicity on the basis of sequence.** On the other hand, as we learn more about pathogen evolution from our studies of nature, we also learn the rules for eventually detecting a pattern of sequences that could be used for monitoring organisms from the wild or those synthesized in the laboratory (see Box 2.7).

The Parallels in the Evolution of Pathogenicity in the Large Viruses

A virus' genes are associated with its ability to replicate and persist in a specific host cell. However, viral pathogenicity is as dependent on the biological context of the host as it is on the viral agent. Rabies can be carried in a bat harmlessly, but is almost always lethal for an untreated, infected human. An infected human is a biological dead end for both the virus and the host. The evolution of pathogenicity in bacteria, viruses, and eukaryotic pathogens involved horizontal gene transfer. For example, the evolution of the large nuclear and cytoplasmic DNA viruses that gave rise to poxviruses was dominated by

BOX 2.7

- ***Homolog.*** A gene related to a second gene by descent from a common ancestral DNA sequence.
- ***Ortholog.*** A gene related to a second gene by descent from a common ancestral sequence and located in different species. Often retain the same function. Genes in different species that evolved from a common ancestral gene by speciation. Often, orthologs retain the same function in the course of evolution. Identification of orthologs is critical for reliable prediction of gene function in newly sequenced genomes.
- ***Paralog.*** Genes related to a second gene by duplication within a genome. May or may not retain the ancestral function. (Paralogs may acquire new functions in the course of evolution, even if they are related to the original one, whereas orthologs often retain the same function.)
- ***Speciation.*** The origin of a new species. As part of this process the new species acquires some barrier to genetic exchange with the parent species.

expansion of paralogous gene families and acquisition—by horizontal gene transfer—of numerous genes from early eukaryotic hosts, other viruses and, rarely, bacteria. Comparative genomic studies of poxviruses support a monophyletic origin from a distant ancestral virus on the basis of the presence of 41 homologous genes common to 16 distantly related viruses that have with large DNA genomes (about 150 to 900-kbp predicted protein-coding genes) (Iyer, Balaji et al. 2006). Those 41 genes are a subset of 90 highly conserved core genes (about 100-kbp) common to all poxviruses, which are at the center of the genome, which encode proteins necessary for replication, transcription, assembly of capsid and the acquisition of a membrane (Upton, Slack et al. 2003; Lefkowitz, Wang et al. 2006). The genes at the flanks of the genome (about 60-100 genes) evolve at a rate distinct from that of the essential genes of the central core, and are more often involved with the mitigation or modulation of cellular processes and whole organism responses to infection, (such as apoptosis, ubiquitin signaling, and cytokine signaling) (Moss 2007). Although the progenitor of poxviruses evolved by gene gain from an ancestor common to large nuclear and cytoplasmic DNA viruses, poxvirus species adapted to new hosts through gene loss, keeping the genes that were necessary to parasitize one particular environmental niche successfully (Upton, Slack et al. 2003; Odom, Curtis Hendrickson et al. 2009). In the case of poxvirus pathogens, variola and molluscum contagiosum viruses, this resulted in a narrowing of the host to only humans. As noted above, evolution through gene loss is also carried out by free-living or facultatively parasitic bacteria that adapt to a more dependent association with a host. One study suggests that variola virus diverged from an ancestral poxvirus of African rodents either about 16,000 or 68,000 years ago, depending on the historical record used to calibrate the molecular clock (Li,

Carroll et al. 2007). During that process, the genome maintained its conserved, core gene complement but lost a unique pattern of virulence genes through accumulated mutations or gene deletions to reach its current size of about 186-kbp (Esposito, Sammons et al. 2006). A similar process occurred in the other orthopoxviruses; each species differed in the number and types of virulence genes that were lost except for cowpox viruses (Lefkowitz, Wang et al. 2006). Of the known orthopoxviruses, cowpox virus has the largest genome (224-kbp) and has orthologues of all other genes found in all other orthopoxvirus species (Gubser, Hue et al. 2004; Lefkowitz, Wang et al. 2006). An ancestral poxvirus, similar in size to cowpox virus, is hypothesized as the progenitor of orthopoxviruses (Gubser, Hue et al. 2004). Given that evolutionary history, it is not clear whether the loss of genetic information was a consequence of the lack of requirement of a gene for propagation in a new host, a necessary prerequisite for propagation in a new host, or both. However, we have available the gene sequences from this entire array of possibilities and we can identify those genes that are part of a common core of the poxviruses and those with more specificity for a particular host.

Nature provides an interesting lesson about the value of sequence identity for predicting pathogenicity. The poxviruses, molluscum contagiosum, and variola, have evolved to become successful human pathogens, but in dramatically different ways. Variola virus is a respiratory pathogen that causes a systemic disease with high mortality rate in humans regardless of age or sex. Molluscum contagiosum (caused by *Molluscipoxvirus* genera) is a common, benign infection of the skin of children and sexually active adults, but it can be a frequent and serious opportunistic infection of immunosupressed patients. [Molluscum contagiosum viruses are more highly restricted in its tissue tropism than variola virus and replicates only in the human keratinocyte (Buller and Palumbo 1991).] Molluscum contagiosum and variola virus share a similar genome structure with a central, conserved core of orthologous genes, and a unique pattern of flanking genes. **The evolution of two distinct poxvirus pathogens that cause dissimilar human disease from a common ancestor through gene loss suggests that there can be multiple genetic pathways to becoming a pathogen for the same host.** Furthermore, inasmuch as the genomes of molluscum contagiosum and variola viruses are distantly related with 57 percent nucleotide identity (on the basis of DNA polymerase), the evolutionary path to a human poxvirus pathogen does not necessarily demand a high degree of genome sequence identity. Even with the closely related orthopoxviruses that share 96 percent nucleotide sequence identity, phylogenetic analysis cannot predict human pathogenic potential (see Box 2.8).

Evolution of Plant Pathogens in Human-Managed Ecosystems

In evolutionary terms, human-managed artificial agricultural ecosystems are a recent agricultural practice that is highly vulnerable to disease outbreaks.

BOX 2.8

Evolutionary relationships, and therefore biological potential, may be inferred through the phylogenetic analysis of orthopoxvirus genomes because the genomic sequence is ultimately the primary genetic map of the species. Examination of phylogenetic predictions based on multiple nucleic acid sequence alignments of the conserved, core genomic regions of isolates of representative orthopoxvirus species found that monkeypox, ectromelia and cowpox viruses (strain Brighton Red) do not group closely with any other orthopoxvirus species, whereas variola, camelpox and taterapox viruses form a more closely related subgroup (Gubser, Hue et al. 2004; Esposito, Sammons et al. 2006). On the basis of that phylogenetic analysis, it would be predicted that if any orthopoxvirus shared biological properties with variola virus and caused human disease, it would be camelpox virus or taterapox virus. Instead, the more distantly related monkeypox virus causes severe human disease that is almost undistinguishable from smallpox, whereas taterapox and camelpox viruses have not been documented to cause any human disease (Damon 2007). Monkeypox and variola viruses together have at least 33 orthopoxvirus virulence genes where a function has been determined for the actual gene or for an orthologue in another orthopoxvirus species. By sequence comparison, both viruses appear to have 21 functional genes in common; of the remaining 12 genes, variola virus lacks 9 and monkeypox lacks 3. Those data suggest that are at least two and probably more unique genetic backgrounds are capable of producing a poxvirus that is lethal for humans. The identity of additional human poxvirus pathogens will probably be an experiment of nature and not a result of phylogenetic genomic comparisons.

Agricultural ecosystems are artificial and contain a high density of a genetically uniform plant population, which creates a local atmospheric zone where the climate differs from that of the surrounding area, and promotes a high rate of pathogen reproduction and dispersal (Stukenbrock and McDonald 2008). Such agricultural practice allows virulent pathogen genotypes to adapt to a particular host genotype and to increase rapidly, and thereby generates a degree of host specificity in a short period that rarely occurs in natural ecosystems (see Box 2.9). Examples highlighting the major evolutionary mechanisms by which plant pathogens have emerged as threats to agricultural ecosystems are provided in detail in Appendix K. Those mechanisms take place over varied time scales ranging from short to thousands of years.

Interactions of Infectious Agents with the Host

The interplay of expression of genomes of a host and a microorganism dictate the outcome of the interaction. Microbial and host interactions can be characterized as stable, dead-end, evolving, or resistant. The most common

BOX 2.9
Evolutionary Mechanism of Plant Pathogen Emergence

Host-tracking

Host-tracking refers to co-evolution of a pathogen with its host during the process of host domestication, which includes the formation of a specific agro-ecological system. It includes the selection and cultivation of desirable host genotypes, simultaneous selection for pathogen genotypes that are adapted to the selected individuals and for the agro-ecological conditions at the time of the process. The process takes about ten thousand years, and pathogen and host share the same center of origin.

Host jump

Host jump is a process through which a new pathogen emerges in a host species that is genetically distant from the original plant host (for example, from another class or order). The geographic origin of the host does not always correspond with the geographic origin of the pathogen as observed in host shift.

Host shift

Host shift is a process in which a new pathogen emerges by adaptation to a new host that is a close relative of the former host (for example, shifting from a wild crop to the new domesticated selection or variety of the crop). The process may take less than 500 years or as much as several thousand years and the pathogen and the host do not always originate in the same center of origin (Stukenbrock and McDonald 2008).

interaction is the resistant host-microbe interaction that leads to no or minimal amplification of the infectious agent. An evolving microbe-host relationship characterizes the spread of a microorganism from the same or a closely related species (as in the West Nile virus introduction into the Americas in 1999). A stable microbe-host interaction results in survival of both the microorganism and host on a population basis (for example, variola virus, polio virus or *M. tuberculosis* in humans and plum pox virus in some stone fruit species). In such cases, the host species acts as a reservoir host and it is necessary and sufficient for completion of the natural life cycle of the infectious agent. The resulting disease can be of varied severity as long as transmissibility of the pathogen is ensured. In general, those interactions evolve toward a state of less pathogenicity in the host, while preserving transmissibility. Stable interactions can also include infection of more than one host species with the same microorganism. For example, influenza A virus, Rift Valley fever virus, *B. anthracis*, tomato spotted wilt virus, cucumber mosaic virus, and turnip mosaic virus are all capable of propagation in a variety of species, and some viruses, such as Rift Valley fever virus, can replicate efficiently in insects as well as mammals. Dead-end interactions often result in severe, fulminant disease involving infections of an incidental host species that is not needed for maintenance of the natural life cycle of the pathogen (for example, *B. anthracis*, dengue virus, Nipah virus, and Ebola virus infections of humans, and *citrus tristeza* virus and *Phytophthora infestans* infections in citrus grafted onto sour orange and potatoes, respectively). The infectious agents originate in other vertebrate species or are carried by arthropods that cycle between insects and vertebrates or between insects and plants. In some circumstances, a "dead-end" infection can give rise to an emerging infection as was the case for HIV, Rift Valley fever virus, and SARS-CoV in humans and citrus tristeza virus and *P. infectans* infections mentioned above. The majority of the human pathogens found on the Select Agent list cause dead-end interactions.

The outcome of a microbe-host encounter is based on interactions at the molecular and cellular level that take place over time. For certain viruses, a productive infection is determined by specific receptors that need to be engaged for virus binding and entry (for example, sialic acid and angiotensin-converting enzyme 2 for influenza A virus and SARS-CoV, respectively) the availability of intracellular complementing factors needed for efficient replication and the ability to manipulate intracellular antiviral signaling pathways (for example, interferon, pattern recognition receptors, apoptosis, and autophagy), and the adaptive immune response. In the case of Gram-negative bacteria, a productive infection may be initiated by adhesion through fimbriae (for example, enterotoxigenic *E. coli*); in the case of Gram-positive bacteria, it may be initiated through cell wall-anchored proteins (for example, microbial surface components recognizing adhesive matrix molecules). Additional virulence factors are needed to counter the innate and adaptive immune response. In the case of plant pathogens, a productive infection is determined in part by success in bridging the plant basal defense or innate immunity system though the expression of countermeasures.

Appendix K, presents examples of various factors that affect microbe-host interactions, and our lack of understanding of the basis of a pathogen's ability to infect a host or multiple hosts even if we know the genomic sequence of the pathogen and, in the case of human infections, the host. A key factor in the outcome of the microbe-host interaction is the effectiveness of the host innate and adaptive immune response in the face of sophisticated and redundant microbial countermeasures, some of which are conserved in bacterial pathogens that infect plants and animals. Another important factor is selection pressure, which can be manifested in physiological, epigenetic and/or genetic changes in a pathogen in response to the innate or adaptive immune system, the absence of an adaptive immune response, or changes in the microbiome (Box 2.10).

We are just beginning to understand the significance of the microbiome for human health. Microbial interactions may determine whether a would-be pathogen acquires increased virulence or transmissibility, and whether an infec-

BOX 2.10
Bacterial Super Infection Following Influenza A Virus Infection

Infection of the upper respiratory tract with such a virus as Influenza A virus predisposes the host to superinfection with bacterial pathogens, including *Staphylococcus aureus* and *Streptococcus pneumoniae*. The massive host innate response to Influenza A is focused on elimination of the virus from the respiratory tract. The overwhelming inflammatory response directed against the intracellular viral pathogen provides a perfect opportunity for a member of the normal respiratory flora (such as *S. aureus* or *S. pneumoniae*) to establish itself deeper in the respiratory tract, where it can cause bronchitis and/or pneumonia. The signals in the bacteria that cause them to transition from commensal organism to pathogen are not entirely known although the complex regulatory networks are slowly being identified and characterized. Once the bacteria migrate to their new niche, a plethora of virulence factors are produced. The host innate response to the bacteria is impaired because the overwhelming response to the virus depletes the neutrophils, macrophages, and dendritic cells and the antimicrobial factors produced by them. As a result, the secondary bacterial infection is often far more life-threatening than the initial viral infection. *S. pneumoniae* has emerged as the most common cause of secondary bacterial pneumonia in the current H1N1 Influenza A outbreak, although *S. aureus*, both methicillin-resistant and methicillin-sensitive, is the second most commonly isolated organism in post-influenza bacterial pneumonia.

tion will result in disease. Thus, the individual microorganism, microbe-host interactions, and the environmental context must be considered in assessing an organism or a gene sequence as a potential threat to human health.

THE SPECIAL CASE OF SYNTHETIC BIOLOGY

Innovations in the chemical synthesis of DNA have led to dramatic improvements—the DNA can be longer, of higher quality, and less expensive per base pair—since the synthesis of the first copy of the 75-base pair $tRNA^{Ala}$ in the early 1960s (Agarwal, Buchi et al. 1970). DNA synthesis on solid supports combined with phosphoramidite nucleosides allowed the synthesis of 2.7 kb plasmid DNA, an infectious 7.5 kb poliovirus genome, a 32 kb bat coronavirus (HKU3) that was the precursor to the SARS-CoV epidemic, and the first complete synthesis of a 582 kb artificial bacterial genome. Most recently, a synthetic bacterial genome has been "booted"[6] into an autonomous life-form, so the artificial bacterial genomes are self-perpetuating (Gibson, Glass et al. 2010).

[6]Synthetic biologists have adopted this terminology from computer science, in which it means "to start (a computer) by loading an operating system from a disk. " In the present case it refers to starting an organism from a genome.

It is clear that the price of DNA synthesis has steadily decreased and a cursory survey of commercial suppliers show a cost of about $0.39-0.50/base—a dramatic reduction from synthesis costs commonly seen in the early history, when industry costs of about $5-10.00/base in the early 2000s were common. New optical deprotection chemistries and microfluidic technologies that allow programmable synthesis of hundreds of thousands of oligomers in parallel with fairly high fidelity seem poised to revolutionize inexpensive synthesis. With those and current multiplex technologies, it seems likely that future costs will approach $0.03/base (commercial costs); additional costs will be associated with gene assembly, quality control and other manufacturing issues. In the near future, as gene synthesis approaches $0.10-0.20/base, synthesis will replace most traditional recombinant DNA methods and allow the ready design and synthesis of new gene circuits and biological processes.

The design and testing of artificial biological systems and understanding of functional interactions are key objectives of synthetic biology. The benefits and broad availability of affordable gene synthesis are expected to foster rapid response platform technologies for producing candidate vaccines and therapeutics to address biothreat agents and newly emerging infectious diseases. It will allow more effective diagnostic platform design and basic inquiry into fundamental biological mechanisms, including pathogenesis and pathogen-host interactions. Gene synthesis will assist in designing and testing complex biological systems to fulfill specific purposes ranging from biofuel and petrochemical production, to genetically engineered foods, virus batteries, solar cells and energy systems, and the manufacture of new medicines. **Thus, when considering steps that aim to prevent the misuse of the technology, we should also recognize the dramatic impact and potential that synthetic biology offers to the future economic growth, competitiveness and viability of the U.S. biotechnology industry.**

In general, the discipline uses either natural cellular components and systems to construct new biological processes (the top-down approach) or the generation of unique biological and/or chemical systems that have novel properties and are designed to mimic living systems (the bottom-up approach). Most investigators have used the former approach because foundational understanding for *de novo* biological design (for example, protein structure, protein-protein interaction, genetic regulation) is in complete. One controversy is that unanticipated outcomes may occur when engineered organisms reproduce, evolve, and interact with the environment. Another concern is the deliberate misuse of the technology to design and construct new pathogens, either by engineering in components that resist current vaccine or therapeutic interventions, or by altering pathogenesis by blending in virulence genes from alterative pathogens, or by the *de novo* design of new pathogens. **Understanding of the complex genetic and protein networks that regulate replication and disease is substantially lim-**

ited, so for the near future, synthesis can for the most part only copy, emulate, or recreate existing gene sets that have been designed by nature.

Top-Down Approach

Since 1980, a standard set of recombinant DNA techniques has been developed that allows the cloning of full length DNA copies of most DNA and RNA virus genomes. In parallel, highly efficient strategies have been developed for "booting" infectivity from the DNA genomes and then recover infectious virus, including recombinant viruses that contain design modifications (for example, mutations, new regulatory networks, and gene insertions or deletions). There are proven strategies exist for reconstituting most of the viral Select Agent and category A-C biodefense pathogens from full-length DNA genomes; however, infectious clones have not been constructed for many Select Agent viruses on these lists. Developing infectious full-length DNAs is traditionally a time-consuming and uncertain process; single nucleotide changes can destroy genome infectivity. In general, genome size is directly proportional to the difficulty of generating an infectious molecular clone because of issues associated with genome and vector stability, sequence accuracy, and technical challenges in manipulating and recovering large genomes. For example, substantial technical sophistication, targeted expertise, and practice are required to reproducibly "boot" genome infectivity of large RNA genomes (for example, coronavirus, Ebola virus and influenza virus), DNA genomes (such as poxvirus) and bacterial genomes. In contrast, small RNA and DNA genomes are much easier to manipulate and recover. The number of people capable of working with these systems is increasing on a daily basis. For example, students at Johns Hopkins University recently worked collaboratively and designed, synthesized and recovered a 280 kb yeast chromosome (Dymond, Scheifele et al. 2009).

Synthetic biology will alter the standard approaches for reconstructing full length infectious DNA genomes of most viruses and computer based genome design will probably become the norm in the near future. **It is clear that gene and genome synthesis will allow for synthetic reconstruction of many highly pathogenic human, animal and plant virus genomes, and thereby, removing one major limit to biological warfare and terrorism: availability.** There are similar concerns regarding synthetic bacterial genomes, which are on a longer time horizon. Assuming that a cost of $0.10/base is achievable in the near future, the synthesis of the genomes of most "agents of concern" will be readily affordable (RNA genome: about 7.32kb for $700.00-3200.00; DNA genome: about 150-300kb for $15,000-20,000), while half a million base pair bacterial genomes will cost about $50,000 U.S. dollars. Not only is the instrumentation affordable, highly portable, and globally available, but there are commercial vendors on every continent, so it would be difficult or impossible to track and police nefarious intent. **Thus, it is possible that no virus (or microorganism) can ever be considered extinct (for example, poliovirus, 1918 influenza, smallpox, reconstructed extinct retroviruses, wooly mammoth viruses, etc.) as long as basic sequence information is available to support its synthetic reconstruction. The considerable concern surrounding synthetic DNA technology for "dual-use" potential is understandable (NRC 2004).**

In general, dual-use concerns include the use of synthetic biology to deliberately host-shift pathogenic microorganisms, to engineer drug or vaccine resistance, or to alter virulence potentials. As noted throughout this chapter, the list of "virulence genes" that have defined biological properties is growing at a considerable rate, and this fuels concerns that virulence is a readily malleable trait. Limited research has focused on the potential of synthetic genome design to enhance viral pathogenesis. With an existing genome as a chassis (from either a pathogenic or a nonpathogenic virus), it is certain that virulence genes from DNA and RNA viruses can easily be introduced into recombinant genomes in an attempt to alter the pathogenic potential of the chassis genome. The capability of using standard recombinant DNA techniques for that purpose on a more limited scale has existed for about 30 years. Moreover, we can imagine synthetic killer viruses that destroy civilization or that cause significant morbidity and mortality—a common topic in cinema.

We know how to synthesize such imaginary "doomsday scenario viruses," but how well the blended genomes will perform in human populations is unknown. For example, expressing the influenza virus NS1 type 1 interferon antagonist gene in SARS-CoV is simple with a top-down engineering approach, but the pathogenic properties of the chimera is difficult to predict. That is because SARS-CoV encodes at least six other interferon antagonist genes, and this raises the question of the ability of NS1 to offer considerable improvements in the pathogenic and innate immune antagonism capacity of the coronavirus. Moreover, most viral proteins form complex interaction networks that are essential for regulating efficient virus growth and virulence. Removing or introducing new potential interaction partners will most likely adversely affect virus-virus and virus-host interaction networks and thus influence pathogenesis outcomes in unanticipated ways. In addition, dramatically altering the genome content of most RNA viruses by inserting genes could easily attenuate virulence, probably by affecting global gene expression or by altering basic RNA structure and genome packaging and release. In spite of those limitations, insertion of the IL-4 gene (cloned from mus musculus domesticus, the common mouse) into murine poxviruses (such as ectromelia virus) or insertion of the SARS-CoV ORF6 gene into mouse hepatitis virus enhanced pathogenesis in mice, and the influenza virus NS1 gene enhanced Newcastle Disease virus replication in human cells. It is also clear that synthetically designed chimeric viruses would elicit fear in exposed populations, regardless of the actual pathogenic outcomes associated with its intentional release. It might be prudent to shift the focus away from preventing dual-use proliferation to preparing for it by developing new platforms for rapid vaccine and therapeutic design and stockpiling these reagents against future bioterrorist attacks.

It is important to note that many "non-Select Agent" human viruses and microorganisms are extremely pathogenic and encode virulence determinants that could be blended into other genome chassis or enhanced by introducing new virulence determinants. **Thus, one danger in creating lists is that it focuses resources on a select subset of human pathogens and ignores the broader innovation in genome design and gene function that exists throughout the larger microbial world** (see Figure 2.2).

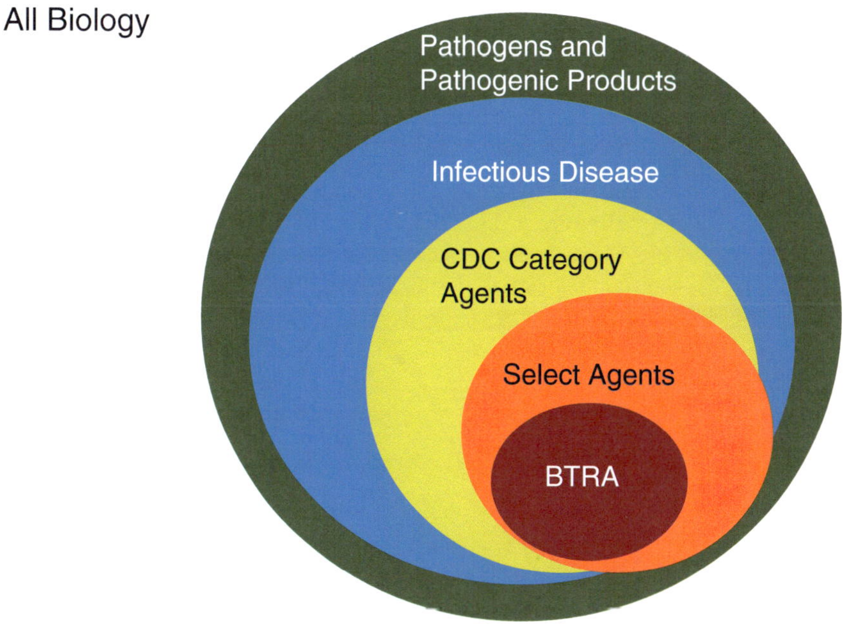

FIGURE 2.2 The universe of potential genes and sequences that could be drawn upon to create a biological weapon involves all biology. From "All Biology," some pathogens and pathogenic products such as toxins, venoms, and others may be known to cause disease or death in humans, animals or plants. In the context of human health, some are recognized as causes of infectious diseases and are reasonably well characterized. Among all infectious diseases, some are further classified as Category A, B, or C pathogens by CDC or NIH, and they may or may not be assigned a biosafety level of laboratory containment (BSL-1, -2, -3, or -4) in the BMBL. Of these, some are designated Select Agents, and a few are prioritized under the DHS Bioterrorism Risk Assessment.

Bottom-Up Approach

The most commonly cited concern is that synthetic biology will allow the *de novo* design and creation of new life forms, such as synthetically designed killer viruses and microorganisms, either from scratch (which is extremely unlikely) or through the combination of existing gene sets from multiple viruses (which is more likely). Humans are surrounded by large numbers of intentionally genetically modified organisms (for example, domesticated corn from small grass teosinte and domesticated microorganisms used in beer, wine, cheese, and yogurt production) that were generated by breeding, artificial selection or, most recently, direct genetic engineering. In fact, 30 years of genetic manipulation with recombinant DNA technology in thousands of laboratories around the world has not resulted in any human-targeted super virus or microorganism. Although that does not preclude it in the future, it raises the question of whether the degree of concern is exaggerated. Will synthetic designer genomes be more dangerous to society than other highly pathogenic microorganisms that already exist in nature, such as measles (about 150,000 deaths per year), HIV (about 2 million deaths per year), influenza and RSV, respiratory syncytial virus, (about 0.5-1 million excess deaths per year), organisms that cause diarrheal diseases (more than 2 million deaths per year), or malaria (more than 1 million deaths per year)?

The scientific community does not have sufficient knowledge to create a novel, viable life form, even a virus, from the bottom up. Designing an infectious viral genome *de novo* by sequence requires the accurate prediction of protein structure and function, the design of protein-protein interactions and protein machines, all of which must produce progeny virions efficiently in an order of magnitude more complex host cell. If we cannot predict protein structure and function on the basis of sequences with any accuracy, how can we design and synthesize novel viruses that will replicate, regardless of their disease potential?

Alternatively, *de novo* design could focus on existing gene sets by emulating or copying known functions. That is also exceptionally difficult. First, entry requires specific interaction between viral and cellular components, including the deliberate orchestration of a series of sequential conformational changes that mediate docking and entry of the virus genome (particle) into the cell. Regulation of entry often involves co-receptors and other host factors like proteases to regulate the entry process. While the process is clear, the ability to design these highly regulated systems *de novo* is in its infancy. Second, a functional replication complex is needed to replicate the viral genome. Replication complexes are complex multicomponent protein machines (such as, viral and perhaps cellular proteins) that specifically engage and replicate nucleic acid. The formation of a replication complex includes tailored protein-protein and protein-nucleic acid interaction networks that are not known and cannot be predicted or engineered with existing technology. Pathogenesis involves a regulated set of virus-host interaction networks that influences host responses that antagonize or potentiate disease; these networks are poorly understood and cannot be designed *de novo*. Virulence genes work together, and the levels of expression that permit virus persistence or spread and transmission, depending on the replication mechanism, are highly regulated. Finally, efficient virus egress from the cell usually involves discrete cellular and viral constituents. The protein-protein interactions, the regulation of the components involved in efficient release, and the design of *de novo* systems are beyond the capacity of the scientific community. **The level of abstraction required to piece together a new life from defined parts is difficult enough—it is a misconception that a viable *de novo* microorganism can be designed today directly from sequences and a pool of uncoupled gene parts—it would be even more difficult to predict the virulence of such a microorganism if it were assembled and recovered.**

Synthetic Biology—Summary

Synthetic biology offers considerable opportunity to improve human health and solve planetary problems in energy, food production, medicine, and public health. However, dual-use applications are of legitimate concern; new synthetic DNA technology alters the old paradigms that regulate pathogen bioavailability, but it is the traditional top-down approaches[7] that present the greatest threat to altering virus virulence and pathogenesis. Bottom-up approaches remain extraordinarily complex, and it seems unlikely that sufficient vision and understanding exist to design and recover a true human pathogen using this approach. The time frame for the realization of the synthetic biology revolution is impossible to predict and will depend on whether it remains unfettered and on the support that it receives; undoubtedly, however, it will be a reality within the lifetime of most who read this document.

WHAT CAN CURRENTLY BE PREDICTED FROM SEQUENCE ABOUT THE IDENTIFICATION OF PATHOGENIC MICROORGANISMS, INCLUDING SELECT AGENTS?

It is abundantly clear that the use of sequence alone to predict a naturally occurring or synthetic pathogenic microorganism accurately is infeasible—sequence cannot provide biological context. Whether synthetic or natural, a sequence by itself has no biological properties. However, a remarkable amount of information can be deduced from a sequence of genes or proteins. To begin with, it is clear that all the microorganisms that we deal with can be divided into relatively well-defined general groups. For example, among the groups of

[7]See Chapter 3 for more discussion of the relative threats presented by these approaches.

bacteria, viruses, and eukaryotic pathogens a core of conserved genes of known and hypothetical function make up the unique set of properties that permit a specific group of microorganisms to occupy, persist, and replicate in a particular biological niche, perform certain metabolic functions, and interact with other living things. That core of genes, which generally constitute about three-fourths of the genes specific to a microbial group, is the basis of much of contemporary taxonomy. A remaining set of genes is concerned with specialization. For our purposes, the specialization is pathogenicity and virulence. We can confidently "predict" the sequence of most of the important microbial toxins or at least suspect, on the basis of the structure of a protein or the deduced catalytic site decoded from a nucleic acid sequence, that a toxin might be encoded therein.

The common features and themes of many bacterial virulence genes are known. Just as we can tell from the sequence of "core" genes that we are dealing with a bacterium of the genus *Yersinia* or a poxvirus, we can accurately say whether there is a precise sequence of a known virulence gene associated with the plague bacillus or with smallpox. It might be difficult to identify the precise origin of a microbial sequence to a specific Select Agent, but it is surely less difficult to say with some accuracy that a particular sequence is related to a known virulence determinant and to a class of virulence genes with known function. The fact is that the availability of complete microbial genomes, as well as the genomes of their hosts, has made it possible to combine genetic and molecular methods not only to identify microbial genes that are expressed only in certain animal hosts, including humans, but also to determine genes that are essential for pathogenicity using global screening methods. The methods themselves depend on gene sequencing technology. Such lists of genes exist today for *Salmonella*, *Mycobacteria*, *Yersinia*, and the pneumococcus; we believe that they probably exist for most Select Agents in at least a surrogate animal model. Where such lists do not exist, it is relatively easy to determine them experimentally.

The utility of sequence information depends on a number of factors. In some circumstances, the quality of a sequence may be a factor, for example, if it comes from a small sample from an infected individual. The length of a sequence is, of course, also a factor. If a sequence is relatively short, it may be of no immediate consequence unless it can be shown to be part of a much larger assembly of genes. What would be required to determine that a sequence came from a known Select Agent? On the one hand, if the sequence were from a known virulence gene or from a specific (as opposed to core) region of the genome, one could say with a good deal of confidence that it was from that organism. On the other hand, if the sequence is from a core common to pathogenic and non-pathogenic microorganisms, as is common for enteric bacteria (and others), the information is less useful, although it is of value from a surveillance standpoint to eliminate suspicion. Difficulty would arise if a sequence were from a common set of core genes—let us say a core common to the *Bacillus* group—and there were sequences that resembled the plasmid-mediated toxin of *B. anthracis* but they were clearly distinct from the known anthrax virulence determinant. That might heighten the suspicion. However, the level of concern would depend on the biological context. If the sequence were derived from a patient who had sepsis, one would be appropriately alarmed. If the sample were from the soil, the level of suspicion would be far less because soil is a natural habitat for members of the genus *Bacillus* and they are harmless bacteria that have plasmids encoding related toxin molecules. If the sample were submitted to a DNA synthesis company, there would be cause to learn more about its source and about the reason the DNA was being synthesized.[8] **There is no fool-proof method for predicting whether a sequence will have the biological properties of a Select Agent (such as pathogenicity), but the common sense use of the considerable amount of sequence data we now have, combined with advances in understanding of microbe-host interactions, does provide for a mechanism that is practical and sufficiently flexible to provide guidance about the potential biological consequences of a DNA sequence from nature or from the biochemist's bench.**

None of this is meant to trivialize the difficulty of understanding the biological importance and function of microbial determinants of pathogenicity or the complexity of microbe-host interactions. Although we have the complete sequence of many of the most feared microbial pathogens and of their host animals or plants, our attempts to devise novel vaccines or therapeutic agents have been difficult at best. If genomics has taught us anything at all about microbe-host interaction, it is just how intimate and intertwined the biology of the two participants may be. That is not always obvious if one thinks about the ravages of the black death of the 14th century or the horror of Ebola infection. Those are the nightmare of those who deal with bioterror, but the most efficient killers of the world are still infectious agents that have a longer-lasting intimate association with humanity. It is not easy to develop antiviral drugs when the interactions of the viral components and the human host are so closely intertwined that inhibiting the invader may mean poisoning the host. Nor is it easy to develop vaccines against many bacterial pathogens that have evolved to interact in subtle specific ways with the biology of the human immune system. Continued investment in genomics, bioinformatics and basic research on infectious diseases of humans, animals and plants, will help us develop new ways to control diseases that plague our planet.

Nevertheless, it is our judgment that the sequence information now available, when combined with that being obtained on a daily basis and deposited in public data banks, can be effectively used to provide a new approach to identifying and monitoring microorganisms of interest much more effectively than the

[8]A common question in this regard is, who should carry the burden—those submitting the sequence for synthesis, the synthesis companies, or government?

construction of rigid lists that reflect biological reality poorly and therefore can provide only limited utility for national security. **Thus, in Chapter 3 we discuss how the Select Agent list might be clarified by sequence-based classification to better circumscribe the taxonomic distinctions blurred by natural and synthetic variation and modification. Moreover, we present a "yellow flag" biosafety system; this approach is not regulatory and therefore could provide sequence information relevant to biosecurity in a more dynamic and timely manner than the Select Agent Regulations.**

3

A Proposal for Consideration: Sequence-Based Classification of Select Agents

The previous chapter concluded that the primary answer to our charge is no—*prediction* of Select Agent status by genome sequence analysis is not feasible.

First, in Chapter 1, the committee found that prediction of Select Agent status is not possible, because Select Agent status involves economic, historical, and policy considerations beyond the biological properties of the organism encoded in its genome.

Second, the committee finds that even if it were possible to assign Select Agent status solely on the basis of genome-encoded biological properties, the answer would remain no. Chapter 2 described why accurate prediction of an organism's pathogenicity from its genome sequence is not possible now, and will not be feasible in the foreseeable future—certainly not at the level of accuracy appropriate for statutory regulations. Reliable prediction of the hazardous properties of pathogens from their genome sequence alone will require an extraordinarily detailed understanding of host, pathogen, and environment interactions integrated at the systems, organism, population, and ecosystem levels. **For the foreseeable future, the only reliable predictor of the hazard posed by a biological agent is actual experience with that agent.**

The committee was charged to identify, supposing that the answer to these questions would be no, "the scientific advances that would be necessary to permit serious consideration" of such a predictive framework. The committee believes that we are so far from the goal of a predictive framework that it is premature to plan *specific* steps towards a Select Agent regulatory system based on *predictive* genome sequence analysis.

As described in Chapter 2, it is a major goal of all biology to understand how DNA sequence determines the properties of biological systems, ranging

in complexity from single macromolecules to pathways, organisms, populations, and ecosystems. Successes in prediction and design at each subsequent level of complexity in biology as a whole are the relevant milestones to watch for, before we will be able to predict confidently from genome sequence analysis how a designed organism would replicate, interact with a host, evade a host immune system, and spread in a population to cause disease.

The committee's view is that for the specific purposes of the Select Agent Regulations, those general biological milestones should be passively monitored, not actively sought. **A narrow focus on such milestones for the sole purpose of being able to predict what makes Select Agents dangerous may be a distortion of priorities in biology, and may also raise concerns about dual use.** The ability to predict pathogenicity from genome sequence automatically confers the ability to design genome sequences of pathogens.

However, the committee is not satisfied with answering its charge narrowly and in the negative. **The rapidly expanding capabilities of automated gene synthethesis and of synthetic genomics to synthesize and "boot" complete Select Agent genomes means that the Select Agent Regulations do need to be defined in terms of genome sequence analysis, not by the phenotypic properties of an encoded agent.** A Select Agent genome is covered by the Select Agent Regulations whether or not it is ever "booted" into a living agent whose phenotype can be assayed. A DNA synthesis company needs to be able to tell, unambiguously and by sequence alone, if it is being asked by a customer to synthesize the genome of a Select Agent.

That determination would not be a problem if each Select Agent had a unique genome sequence. However, discrete taxonomic nomenclature in biology is already challenged by the great diversity and continuum of organisms observed in natural wild isolates, and **the rapidly expanding ability of synthetic biology to create highly modified variants and chimeras of naturally occurring genomes poses an even greater challenge to taxonomic naming systems.** Select Agent pathogens, like any biological organism, are not defined by a single DNA sequence. Given natural wild variation and the conceivable range of tolerable synthetic variation, a "cloud" of related sequences of similar biological properties are all assigned the same taxonomic name. There may be sequences that are just as closely related but are not Select Agents, including vaccine strains and attenuated research strains that the U.S. government want explicitly to avoid encumbering with the Select Agent Regulations.

In its deliberations, the committee found that it was useful and important to distinguish sequence-based *prediction* of biological properties from sequence-based *classification*. A regulatory system based on *prediction* must be able to recognize that an entirely novel genome sequence (unrelated to any known sequence) encodes a pathogen that should be assigned Select Agent status. A regulatory system based on *classification* "merely" tries to decide whether a sequence is sufficiently similar to that of an already-known, already-named Select Agent to be assigned the same taxonomic name and status. The two concepts are easily confused and sometimes conflated because the state of the art of prediction from biological sequence is generally not based on a physics-based prediction of the molecular structure and function of the parts encoded by a genome, but rather on sequence comparison and classification: If one sequence is similar to another known sequence, it is assumed that they share evolutionary ancestry and have similar biological functions.

This chapter explores the conceptual difference between predictive systems and classification systems and considers the ramifications of using sequence-based classification for the Select Agent Regulations. In a narrow sense, the committee has addressed its charge by explaining that prediction-based systems are not feasible. However, the committee interprets its charge more broadly and in this chapter moves beyond infeasible predictive systems to consider a feasible sequence-based classification system. The reader may want to view this chapter as a proposal for consideration, rather than as a recommendation. However, in the sense that sequence-based classification is conflated with rough and limited prediction in biology, this chapter is the committee's positive and constructive answer to its charge. We discuss how a sequence-based classification system might be used to encompass what we believe are the most technically feasible and likely scenarios whereby synthetic genomics and synthetic biology could be used to construct a hazardous agent with Select Agent properties. A sequence-based classification system would still be based on a discrete list of Select Agents, but could be used to create a pragmatic "brighter line" for deciding whether a new genome sequence should be regarded as one of the existing Select Agents or not.

NOVEL AGENTS: SYNTHETIC GENOMICS AND THE SELECT AGENT REGULATIONS

We need to examine what we mean by a "novel" synthetic agent. **This chapter flows from the strong premise that for the foreseeable future, synthetic pathogens (at least those regulated by the Select Agent Regulations) will be composed largely or entirely of genes derived from existing pathogens.** That premise requires careful consideration. We want to outline the most likely threat scenarios that might be created with synthetic genomics, and we want to consider what we should regulate at the level of possession and transfer of specific agents with regulations like the Select Agents Regulations, as opposed to overarching statutory prohibition of use and development of offensive biological weapons under the Biological Weapons Convention, or as opposed to prudent laboratory biosafety guidelines for handling of any pathogenic organism.

Synthetic genomics poses three main threat scenarios that would allow a "bad actor"[1] to obtain a pathogenic organism suited for use as a weapon:

- The bad actor orders a synthetic DNA (or RNA) genome of a known Select Agent. The exact sequence may be modified, either in an attempt to alter the phenotype of the agent (perhaps to introduce drug resistance, increase pathogeneticity, or alter host range) or in an attempt to circumvent the Select Agent Regulations themselves (by making genetic changes intended to have little or no effect on pathogenicity, but to create ambiguity about the taxonomic classification of the organism). The bad actor then "boots" the synthetic genome into a full organism. We will call such an organism a "*modified* Select Agent" and we will describe its creation as "modification" of an existing organism.
- The bad actor "assembles" a synthetic pathogen by combining parts (genes and regulatory sequences) of known organisms, for example by creating a chimera of two or more viruses, or by attempting to express genes that encode a toxin or mechanism of pathogenesis in an otherwise innocuous "chassis" of a non-Select Agent host, such as any of several commonly used viral vectors. This would include cases in which no individual part originates in a Select Agent.[2] We will call the organism created by this scenario a "*chimera*," and we will describe its creation as "*assembly*" of existing parts.
- The bad actor "designs" a synthetic pathogen by creating entirely new gene sequences—dissimilar to any known pathogen gene sequences in nature. We will call the organism created by such a process a "*de novo*" novel agent, and we will describe its creation as "*design*" (although sequences may be selected from randomized pools by high-throughput in vitro evolution or selection rather than actually designed).

Obviously, there is a continuum among these three scenarios. As more new genes are moved into an existing organism, the line between modification and assembly blurs. Only part of an organism may be designed, and the rest assembled or modified. The important distinction is between genetic modification of an existing organism in essentially its original order, assembly of parts of

---

[1]"Bad actor" is used to mean "an individual or group with nefarious intent." "Individual" or "person" does not convey the full meaning; "terrorist" was viewed as too strong or specific a term. In addition, a "bad actor" may be either an individual or a group.

[2]For example, experiments unexpectedly found that inserting a mouse interleukin-4 (IL-4) gene into ectromelia virus (mousepox) created a highly pathogenic virus against mice. Based on these results, there is concern that a related human virus like vaccinia, engineered to express human IL-4, could become highly pathogenic for humans, though neither vaccinia (the smallpox vaccine virus) nor the IL-4 gene by themselves are Select Agents.

known organisms in new combinations in new orders, and creation of entirely new gene sequences dissimilar to any known pathogen genes.

The three scenarios are ordered by increasing technical difficulty and therefore by decreasing likelihood (see Table 3.1). We want to be sure that we are dealing with the more likely scenarios before worrying inordinately about less likely ones.

The first scenario is the simplest, easiest, and most likely to work in the absence of an expensive research and development program; therefore, it is the most dangerous. Most Select Agent viruses can now be booted from synthetic DNA, some relatively easily (such as small positive-stranded RNA viruses) and some with great difficulty (such as large DNA viruses like variola). As the scope of DNA synthesis increases and the technology becomes commoditized, an increasing number of Select Agents can be reconstituted with a modest level of skill in molecular biology. Synthetic genomes identical to complete Select Agent

TABLE 3.1

	"Modified Select Agent"	**"Chimeric Select Agent"**	**"*de novo* Select Agent"**
Three threat scenarios posed by synthetic genomics:	Created by "modification" of an existing organism.	Created by "assembly" of existing parts.	Created by "design" (or by high-throughput in vitro evolution or selection)
Feasibility of scenario given the state of the art:	Feasible Modification is routine genetic engineering	Possible Assembly is a frequent molecular biology technique as an extension of "modification." Assembly of a radically novel agent is beyond the state of the art.	Improbable Beyond current capabilities; if possible at all, likely to require extensive experimental selection, refinement, and testing in susceptible hosts.
Potential solution:	Sequence-based classification. Anchored around a taxonomic name and a full genome reference sequence, this would define the "space" around each select agent—essentially translating the regulatory language into an operational definition that accommodates the biological complexity.	Gene-sequence-based classification. This would identify individual "parts" of genomes (beginning with select agents), define "space" around each part, and determine which "parts," when assembled, are operationally considered a "complete" genome for the purpose of the Select Agent Regulations	Sequence-based function prediction. Neither design, nor prediction is currently possible. Attempting to design and create a bioweapon is prohibited by BWC and USC Title 18 Section 175(b), among others.

genomes are already clearly covered by the Select Agent Regulations. Whether a modified sequence is still called a Select Agent or not (the question of when the magnitude of modification requires a new discrete name) is a gray area in the current Select Agent Regulations.

The second scenario is plausible, but unlikely to work without an extensive research and development effort. Chimeric assemblies are the leading edge of early successes in "synthetic biology." There are efforts to create standardized "parts"—genes or combinations of genes that have functions that are readily transferable into an organism "chassis" that provides basic biochemical, structural, and replication functions (Kwok 2010). The simplest chimeric pathogens would express a toxin gene in a chassis, but the most dangerous toxin genes are already covered by the Select Agent Regulations regardless of what organism they are in. Chimeric pathogen weapons that evade the current Select Agent Regulations would require the assembly of less obvious pathogenesis mechanisms than expression of known Select Agent toxins. The more such an assembly deviates from a known organism, the less likely it is to work, for all the same reasons that prediction of pathogenicity and other phenotypes from genome sequence is not possible. It is not possible with the current state of knowledge to predict and foresee all detailed interactions of gene products that determine an organism's overall phenotype. It would require experimental characterization in suitable hosts to be sure that a chimeric weapon worked as intended. For some Select Agents, there are no surrogate experimental hosts for characterizing virulence, and the only suitable host for a human pathogen may be a human. Those considerations raise the research and development bar substantially, and expose such a program to existing legal prohibitions other than the Select Agent Regulations. Thus novel chimeric constructions are unlikely as terror weapons,[3] although they are alleged to have been used in national offensive bioweapons research programs.

The third scenario is beyond the capabilities of current biology. There is no example of a designed organism or even of a designed genetic pathway, let alone a designed pathogen. Designing a self-replicating organism that has only to interact with simple molecules in a test tube is one thing, and it is hard enough; designing a pathogen that has to interact with a complicated host, evade its immune system, and be transmissible in the natural environment adds daunting layers of biological complexity. There are very few examples of single protein sequences that have been designed to fold in a particular novel way (Kuhlman, Dantas et al. 2003). These first few modest successes in *de novo* design of single proteins constitute the current state of the art. Design and prediction go hand

[3]Or at least as functionally dangerous pathogens. A release of a chimeric construct of suitably scary-sounding parts could cause societal damage from fear and panic even if it were utterly ineffective as an organism. The number of such imaginable scenarios is enormous—probably beyond the scope of the Select Agent Regulations to regulate effectively without overly impeding beneficial biomedical research. The best countermeasure against "mock" chimeric weapons is likely to be a resilient public health emergency response and communication system.

in hand; our lack of predictive ability in biology means that we cannot now design genomes.[4]

Synthetic biology's use of metaphors like "booting" a genome into a living organism or use of a well-known organism, such as *Escherichia coli* as a "chassis" for hosting synthetic constructs (Lee, Chou et al. 2008) may be misleading about the likelihood of *de novo* design. Among synthetic biologists, this metaphorical language emphasizes the long-term goal of making biological systems as engineerable as computers or machines; but the language also tends to trivialize the complexity of biological systems and the enormous gaps in our understanding of them by making it seem (perhaps especially to non-biologists) that we can already engineer biological systems easily. There are examples in which synthetic genomes have been "booted" into living viruses (Cello et al. 2002) and now even cells (Lartigue et al. 2009; Gibson et al. 2010), but it must be remembered that these experiments have "only" synthesized minor variants of known natural genome sequences. An *E. coli* "chassis" has been used to express genes for a complex synthetic function, such as the engineering of *E. coli* to produce pharmaceutically important natural plant products, such as the antimalarial drug arteminisin (Chang et al. 2007), or bulk chemicals, such as 1,3-propanediol (Bio-PDO, Dupont), a starting point for synthesis of plastics (patent WO/2004/101479), but these efforts have required massive multiyear iterative bioengineering and development processes "just" to move known genes from other organisms into the *E. coli* system and get them to work as desired. All existing (and reasonably foreseeable) uses of synthetic biology involve modification or rearrangement of existing biological components. The entirely *de novo* design of genomes and organisms remains science fiction.

We have distinguished three kinds of novel synthetic organisms because we believe that there is a tendency to imagine nightmare scenarios in which a *de novo* unnamed pathogen, dissimilar to any known pathogen and thus unrecognizable by any sequence comparison protocol, is created deliberately or accidentally with synthetic genomics. **Clearly, a regulatory system like the Select Agent Regulations based on a list of known agents and their genome sequences is not effective for regulating entirely *de novo* agents.** Such a concern seems to have been behind the charge to our committee. If one were worried about prohibiting possession and transfer of *de novo* agents at the point of their

[4]An important exception to the concept that design and prediction go hand in hand is that it is feasible to select *de novo* sequences for particular functions by high-throughput in vitro evolution or selection, and thus make it possible to arrive at functional sequences without designing or understanding them. Although selection and directed evolution have been widely applied at the level of individual RNAs or proteins, no novel genome sequences have been created this way. Given the complexity of the problem (the number of possible DNA sequences rises in proportion to 4^N for a sequence of length N), it is extremely unlikely that complete *de novo* genomes could be selected in a small number of iterations by such procedures. If it were done, it would require a long effort of iterative artificial evolution—expensive and complicated work beyond any current research program in biology.

creation, one would need to develop a forward-looking system that predicts the Select Agent status of a *de novo* genome sequence, rather than a backward-looking system based on a taxonomy of known Select Agents. However, as discussed in Chapter 2, such a prediction based system is not feasible.

We instead take the view that not only is the creation of *de novo* agents the most unlikely scenario, but also that it is not and can not be the purpose of the Select Agent Regulations to regulate *de novo* agents any more than it is their purpose to regulate access to novel emerging diseases in nature. The Select Agent Regulations should not aim to prevent access to all possible pathogens. There is no way to anticipate all possible novel pathogens. The Select Agent Regulations should aim to impede access to the most dangerous *known* pathogens. The best defense against the unquantifiable threat of novel synthetic pathogens is not the Select Agent Regulations, but continued enhancements of the laboratory and clinical biosafety measures that we already have to deal with the real and measurable threat of emerging natural pathogens. When a novel agent emerges, research is initiated to study its mechanism of action, the potential threat that it presents, and its susceptibility to countermeasures. When the novel agent has been shown to meet the criteria for Select Agent regulation—with respect to not only its genome-encoded biological properties, but other medical and policy considerations—its name can be added to the Select Agent list so that future access to it can be regulated.

It is our view that the main biosecurity threat scenarios start with accessibility of proven known pathogens, so it is reasonable for the Select Agent Regulations to be backward-looking and based on a list of known agents. These regulations protect us by restricting the availability of agents that we know from experience to be extraordinarily dangerous and that have a high potential for biowarfare or bioterror.

The foregoing establishes the premise for the remainder of this chapter. **Modified Select Agents, made facile by the commoditization of synthetic genomics, constitute the most important and pressing practical issue related to the Select Agent Regulations.** The taxonomic nomenclature of microorganisms is designed for wild isolates of actual organisms that have observable growth phenotypes, not for non-natural modified sequences that exist only as genomic DNA. A system based on (natural) taxonomic nomenclature does not establish a bright line that is sufficient for clear statutory regulation of possession and transfer of synthetic genomic DNA sequences.

As a specific working example, consider the situation of a DNA synthesis company. A DNA synthesis company is capable of synthesizing complete genomes, and the company (if unregistered) formally violates the Select Agent Regulations if it possesses or transfers a synthetic Select Agent genome.[5] Who

---

[5]Although we use synthetic genome orders as an illustrative example, we must recognize that from the standpoint of impeding bioterror scenarios, there will be myriad ways to get around any

judges whether a DNA sequence constructed by the company is considered to be a Select Agent? How similar to a Select Agent reference sequence does a DNA sequence have to be to still be deemed a Select Agent? Currently, each gene sequence company must define for itself the sequence boundaries around each Select Agent, with only minimal guidance from regulators (DHHS 2009). The companies understandably want increased clarity about what sequences are and are not covered by the Select Agent Regulations. By addressing this issue as an example, we would also deal with a number of other scenarios in which synthetic genomics might be used to create modified Select Agents. And we will also be able to deal with some of the most obvious and likely ways that chimeric agents might be assembled with synthetic biology.

As discussed in the next section, we take the view that the most pressing issues can be treated as a sequence *classification* problem more than a sequence-based *prediction* problem.

CLASSIFICATION IS DISTINCT FROM PREDICTION

Whereas sequence-based *prediction* of the properties of Select Agents will not be feasible in the foreseeable future, sequence-based *classification* can be addressed with current technology. At least at the level of individual gene sequences, there is an extensive literature on methods for automated classification of sequences into operationally defined taxonomies. Cellular organisms are routinely classified taxonomically by using small subunit ribosomal RNA sequence comparisons. Everyday examples of assigning protein sequences into annotated families include databases such as Pfam, Interpro, and TIGRfams. The existing literature does not quite suffice for the problem of sequence-based classification of complete Select Agent genomes. Viruses lack ribosomal RNA, for one thing; for another, we need to think about distinguishing a complete genome from a partial one; and we need to worry about artificially modified and chimeric constructs that would not be constrained to follow the patterns of natural sequence

---

screening procedure used by DNA synthesis companies, ranging from splitting an order into apparently innocuous pieces among multiple companies to using offshore companies that do not adhere to U.S. regulations, to simply not using a DNA synthesis company at all. The technology of DNA synthesis is rapidly being commoditized, and DNA oligonucleotide synthesis machines can already be purchased cheaply from eBay. An ebay.com search on "oligo synthesizer" on 10 October 2009 found a used Applied Biosystems 394 DNA/RNA oligo synthesizer on sale for $8,900 (plus $106.16 shipping within 3-8 business days to a committee member's home in Northern Virginia). With great difficulty and specialized technical skills, genes and even whole genomes can be assembled from individual short oligonucleotides. In much the same vein, a determined bioterrorist can obtain isolates of a Select Agent from the wild. The Select Agent Regulations can only raise the difficulty bar for acquiring cultures of proven highly virulent agents and provide law enforcement with tools to prosecute for possession of variants of such agents; because natural biological organisms are widely available, readily engineered, and increasingly easy to create, it is unrealistic to try to design the Select Agent Regulations to preclude acquisition completely.

evolution. However, the adaptations needed to define complete Select Agent genomes operationally are fairly obvious, as we will discuss.

The computational sequence analysis technologies used for sequence-based classification define "sequence spaces" that circumscribe the known variation of sequences that are considered to belong to a useful name, while excluding the known variation of sequences that are considered to be attached to different names. Therefore, a necessary precondition is to have a number of representative sequences that belong to the desired classification and a number of the most closely related sequences that do not belong.

An important principle of automated classification (also known as "*supervised learning*" methods, in statistical inference) is that given the *known* sequences of things that we want to label as Select Agents and things that we do not want to label as Select Agents, there is *always* a classification scheme that can achieve the desired labeling of known sequences with 100 percent accuracy. The important concern is not that a classification system would misclassify a known sequence, but rather how well a classification system generalizes to correctly classify new sequences that it has not seen before. The existing methods for sequence-based classification of protein and DNA sequences provide a flexible set of software tools for human experts to use to define appropriately generalized sequence spaces on a case-by-case basis using expert knowledge. The methods enable 100 percent classification accuracy on known sequences (essentially by definition—the classification system is defined on the known sequences already labeled with a set of known labels), can be expected to perform reasonably well on new sequences, and are readily updated if and when erroneous classifications occur.

The basic principle of sequence-based classification is simple (Figure 3.1). A sequence is assigned the name of the known group that it is "closest" to, as long as it is also within the range of known variation accepted for that group. If it falls outside the range of known variation of any known named groups, it is operationally declared a "novel" sequence, possibly within some larger and more broadly defined sequence family. The key is the definition of *close*, that is of what *distance* means in comparing sequences; various automated computational procedures differ in the details of their approach.

Sequence-based classification is related to taxonomic (species-level, evolutionary) classification, but the two do not always coincide. They coincide when the desired sequence classification corresponds to slowly evolving traits that are shared amongst a "clade" of evolutionarily related organisms to the exclusion of more distantly related organisms. For example, all variola (smallpox) virus isolates have sequences that are more similar to one another than are variola virus and the most related non-Select Agent orthopoxvirus. Sequence-based classification may differ from evolutionary classification for rapidly or recently evolved traits, particularly if only a small number of crucial changes are involved and the same sequence changes and phenotypes have evolved more

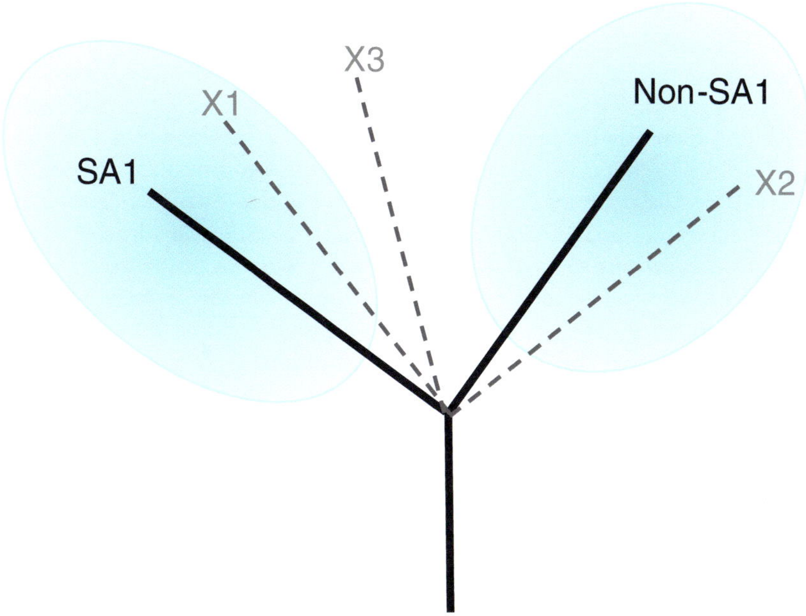

FIGURE 3.1 Profile-Based Classification System—distance.
This graphic shows a tree based on sequence similarity. The Select Agent has high sequence similarity to the non-Select Agent nearest neighbor. The shaded spheres indicate the sequence "space," or known variability, surrounding an agent's taxonomic name.

- X1 sequence is within SA1 "space" and would be classified as Select Agent1.
- X2 sequence is most similar to non-SA1 and would be classified as non-Select Agent1.

IMPORTANT—this is not prediction! X1 may NOT be a pathogen. X2 may be a pathogen. This classification system identifies and clarifies what is subject to Select Agent Regulations. It cannot predict what is or is not a pathogen.

IMPORTANT—classification does NOT designate new Select Agents. It identifies sequences as belonging to a known sequence "space." If the sequence does not belong to an established "space," it is novel.

- X3 is a novel sequence that is more similar to the Select Agent than the non-Select Agent. This would initially be classified as a non-Select Agent. As biological information is acquired, this agent would be evaluated and might be added to the Select Agent list; or it would be considered another near neighbor, non-Select Agent sequence.

IMPORTANT—acquiring sequence data and biological information is needed to define the space around Select Agents and close the gap around novel sequences.

than once in different lineages (this is referred to as convergent evolution). For example, a vaccine strain may differ from a wild-type virus in only a few residues, so a sequence-based classification system might have to focus its definition of the vaccine strain's space on the specific residue identities. Sequence-based classification is strictly operational (in the sense that it can be applied to non-evolved artificial sequences, not just to naturally evolved ones) as opposed to sequence-based taxonomies that are backed by a notion of biological species evolution and evolutionary trees. **A sequence-based classification draws "lines" *(decision boundaries)* that separate desired groups in sequence space; it is, not necessarily constrained by attempts to reconcile the patterns of common evolutionary descent of an organism and its individual genes. That is important, because we are interested in classification of genetically modified organisms and synthetic DNAs that may have no ancestors or no evolutionary history *per se*.**

The "space" defined by a sequence-based classification will usually be large—almost always vastly larger than the space of verifiably viable sequences. That is one reason to use methods that define a space rather than simply making a list of the sequences that we want to include in a group. For example, if we crudely defined the variola virus "sequence space" as everything within 10 DNA substitutions of about a 186,000 nucleotide reference variola sequence, there would be over 10 such sequences.

It is unlikely that all those over 10^{50} sequences would function as a variola virus. Many substitutions would be deleterious or lethal to the virus. Predicting whether and how well each one would function is the realm of sequence-based prediction, whereas in a sequence-based *classification* scheme we do not necessarily care. Rather, we care whether the classifications we assign to viable sequences (or at least to sequences that someone would have a reason to synthesize) are correct. It is not a crucial mistake to misclassify a non-viable hypothetical genome that no one would make. That is a key distinction between prediction and classification. The classification system is not predicting that sequences actually "work" but only that they are closer to those of a known Select Agent than to those of any known non-Select Agents. **Our main concern in using a broadened definition of a Select Agent's sequence space is to balance the need to encompass the most likely modifications and chimeras with the need to avoid the classification of a useful non-Select Agent genome (including those of vaccines and attenuated research strains) as Select Agents.** We need only be sure that we define suitable non-overlapping sequence spaces that capture all known and expected variation in a named Select Agent and the nearest non-Select Agent groups.

The size of the "space" around any given classification will depend on how closely related the nearest competing classifications are. That is important; classification boundaries are arbitrary and operational and are defined by human experts for the particular classification task at hand. There is no single threshold (for percent identity, BLAST score, and so on) that suffices to define the boundary around biological names.[6] The boundaries for a name's sequence space must always be set relative to the nearest neighbors that have different names. Experts might define a tight space around a vaccine strain of one virus, a larger space around the pathogenic Select Agent form of the virus, and several competing "adjacent" spaces of different sizes representing the sequence variation observed in non-Select Agent natural relatives of the virus.

High-throughput automated annotation of individual gene, protein, or RNA sequences as belonging to particular sequence superfamilies, families, and subfamilies is already a robust and routine technology. At the whole (cellular) organism level, sequence-based classification has most often been done by selecting one representative gene sequence (usually small subunit ribosomal RNA) and assuming that the phylogenetic tree of that sequence reflects the phylogenetic tree of the species. More recently, methods have begun to calculate consensus phylogenetic trees on the basis of the simultaneous analysis of multiple genes. Virus taxonomy is somewhat more challenging; viruses lack ribosomal RNA genes or any other universally conserved sequence, and seem even more prone to lateral gene transfer than bacteria. But it is likewise increasingly reliant on genome sequence-based classification that uses key conserved genes in each virus group. With a modest amount of research and development for the specific application, methods like those could be adapted to the problem of sequence-based classification of modified Select Agent genomes, and to some extent to chimeric assemblies.

SYNTHETIC GENOME CLASSIFICATION UNDER THE CURRENT SELECT AGENT REGULATIONS

As noted earlier, the current Select Agents Regulations cover not just culturable organisms, but also naked DNA, including synthetic genomes. The Select Agent Regulations language confers Select Agent status on "nucleic acids that can produce *infectious forms* of any of the Select Agent *viruses*" and "recombinant nucleic acids that encode for the *functional form(s)* of Select Agent toxins" (italics added). In a document interpreting the Select Agent Regulations with regard to synthetic genomics ("Applicability of the Select Agent Regulations to Issues of Synthetic Genomics" see Appendix E), we find a written interpretation that "non-infectious components of Select Agent viruses," including "genomic fragments from Select Agents" and "material from regulated genomes that has been rendered non-infectious" are not covered as Select Agents.

This guidance is talking about using DNA synthesis technology to produce an exact copy of a known Select Agent. That is the most likely application of synthetic genomics; a naturally occurring genome sequence is already known

[6]That is, each Select Agent requires that its own sequence space to be defined.

to function.[7] Almost all Select Agent genomes have now been completely sequenced. The guidance is not attempting to deal with using DNA or synthetic biology to create a novel (modified, chimeric, or *de novo*) "synthetic" agent. The difficulties of dealing with modified, chimeric, and *de novo* synthetic agents constitute the reason for the charge to our committee.

Limiting the nucleic acids language to viruses and toxins reflects the current technical state of the art that synthetic microbial genomes cannot be "booted" into new organisms, whereas many viruses can be, and toxin genes can be expressed in any host organism chassis (such as *E. coli*). The technical barriers to booting microbial genomes into new organisms are starting to fall (Lartigue et al. 2009; Gibson et al. 2010). **Synthetic biology will sooner or later be able to "boot" a wide variety of organisms from synthetic DNA. The Select Agent Regulations language will need to broaden to cover nucleic acids of all Select Agents that can be booted from synthetic genomes.**

The stickier point from the standpoint of sequence analysis and sequence-based classification—and where one enters a grey area of modified synthetic agents, with engineered differences from wild-type Select Agents—is what exactly is meant by such terms as *infectious forms* or *functional forms* or *genomic fragments*. This might mean having a complete set of all the right parts to *attempt* to boot an organism, or it might mean that the resulting organism would actually be expected to *work* as an infectious and/or functional Select Agent. For example, the easiest viruses to boot are the small positive-stranded RNA viruses because the RNA itself is "infectious"; a full-length RNA can simply be synthesized and transfected into a host cell, where it will produce new virus. Researchers refer to the complete RNA transcript as "infectious" in the sense of being complete and sufficient to assemble new virus in laboratory cultured cells, regardless of whether the new virus is "infectious" in the sense of transmitting serious disease to humans or other host organisms. For example, an RNA virus laboratory might synthesize and transfect "infectious" mutant RNAs to study defects in virus replication or maturation in cultured cells, although any resulting viruses might not be harmful to humans.

There is no way for a DNA synthesis company to decide, from sequence alone, whether a variant genome sequence is "infectious." Advances in synthetic genomics make it clear that the Select Agent Regulations definition is underspecified. Drawing a distinction between an apparently complete set of parts to produce a virus and an actual working pathogenic virus and considering the problems posed by synthetic and natural variation bring us back to

[7] A qualification: DNA sequencing has an error rate that ranges from about 1/100 to $1/10^6$ in most cases, depending on the sequencing strategy. Because of DNA sequencing error, the genome sequence in the public databases is not guaranteed to be a functional genome. DNA synthesis technology also has an error rate that makes it non-trivial to construct viable large constructs (Gibson, Glass et al., 2010).

the distinction between classification and prediction. Determining whether a sequence encodes an "infectious form" of a Select Agent is an empirical experimental question and will long remain beyond any foreseeable predictive ability of biology. However, we should be able to use sequence-based classification to establish a reasonable operational definition of the sequence space that circumscribes *complete* agent genomes, as distinct from incomplete genomes or complete genomes of related non-Select Agents.

There is an important distinction between identifying a suspicious "sequence of concern" that might be part of a Select Agent and determining that a genome sequence is "complete" ("infectious") and therefore itself subject to the Select Agent Regulations. The former can be in a gray area, but the latter should be as bright a line as possible. The Department of Health and Human Services (DHHS) has recently published a notice, "Screening Framework Guidance for Synthetic Double-Stranded DNA Providers" (DHHS 2009), that gives guidance to DNA synthesis companies on how to screen orders for possible components of Select Agents. The purpose of the DHHS screening guidance is not to define a complete Select Agent genome for the purposes of regulation but rather to harmonize the ways in which companies should screen customers and sequences and follow up on any suspicious orders with additional questions and scrutiny to be sure that appropriate biosafety procedures are used and regulations are followed. The guidance describes a "best match" procedure for identifying that one or more 200 nucleotide fragments of a synthetic DNA order is more similar to a known Select Agent genome sequence than to any non-Select Agent sequence, by a local alignment search of the entire available nucleotide database (Genbank) at the amino acid coding level. Our committee found the suggested guidance in this Federal Register document to be strong, useful, and well thought out for the purpose of identifying sequences of concern, and (as we will discuss in more detail later) "best match" by percent identity is a reasonable rule-of-thumb sequence analysis procedure. However, the guidance is neither well defined nor stable enough for use in defining Select Agents for regulatory purposes. It relies on non-trivial manual examination of results (Select Agent versus non-Select Agent sequences are not clearly identified in Genbank), and the Genbank database itself is rapidly growing and changing, so results will vary. The guidance also makes no attempt to define a complete Select Agent genome.

The concepts that we develop below expand on the existing synthetic genomics guidance, and we describe how a reproducible objective definition of complete Select Agent genomes might be accomplished to delineate clearly which synthetic DNA constructs are and are not covered by the regulations. We emphasize that the system described is based on classification rather than on prediction. The technologies and scientific knowledge required for this system are available or could be available soon.

CLASSIFICATION DEPENDS ON BOTH GENE CONTENT AND GENETIC DISTANCE

The problem of classifying a genome sequence as a complete Select Agent genome has two dimensions: (a) *content*—how much sequence (how many parts) must be present to distinguish a potentially complete "infectious form" of the Agent from a non-covered "genomic fragment" or "non-infectious component"; and (b) *distance*—how close must the sequences of each of those parts be to an actual Select Agent sequence for the same Select Agent taxonomic name to be assigned to the synthetic organism.

Confusion will arise if content and distance are not distinguished. For example, the language of 18 U.S.C. 175c that the National Science Advisory Board for Biosecurity recommended repealing defines variola virus as "a virus that can cause human smallpox or any derivative of the variola major virus that contains more than 85 percent of the gene sequence of the variola major virus or the variola minor virus."

What does "85 percent of the gene sequence" mean? Does it mean a fragment of 85 percent or more of the full-length variola virus genome? Does it mean a full-length genome of at least 85 percent or greater sequence identity to variola? The NSABB seems to have interpreted it as the latter ("the definition of 'variola virus' in 18 U.S.C. 175c is based on genome sequence similarity"). They point out that this is problematic because "there are many genomes of the variola major virus genome and variola minor genomes that are significantly greater than 85 percent similar to sequences found in related but relatively harmless viruses." With that interpretation, 18 U.S.C 175c outlaws vaccinia virus, the smallpox *vaccine*. The NSABB recommended repealing this problematic language "particularly because the misuse of variola virus is adequately covered by other criminal laws already in place" (NSABB 2006:12), including the Select Agents Regulations and the Biological Weapons Convention. However, as we are seeing, although the language of the Select Agent Regulations avoids making the error of an overprecise and overbroad definition, it could easily be criticized for being underspecified with regard to technical capabilities in synthetic genomics.

We will deal with the issue of content versus distance separately.

USING "PARTS LISTS" TO DEFINE GENE CONTENT

"Content" is the first issue to address. What combinations of "parts" make a complete Select Agent genome? In the genome of a known Select Agent, how many parts can we alter, delete, or substitute before the organism ceases to be dangerous enough to still be considered a Select Agent?

Our state of knowledge will necessarily be partial. It is not feasible to test all possible alterations of a genome sequence experimentally. However, we have clues, from comparative analysis of genes conserved in similar organisms and from experimental studies. We generally have a good idea of what genes are essential (required for viability of the organism under laboratory conditions, such as an RNA polymerase gene in an RNA virus), and some idea of what genes appear to be most critically involved in pathogenesis mechanisms. For the purposes of sequence-based classification, we do not need to have complete knowledge. Partial answers, given the current state of knowledge, suffice for an operational definition of a "complete" Select Agent genome. If an agent actually has 20 genes essential for viability and pathogenesis, and we know only about 10 of them, and we define any genome that contains those 10 genes as a "complete Select Agent genome," our definition is biologically incorrect, but it can suffice as an operational definition.[8]

That is, for an operational definition of a complete Select Agent genome we can define a parts list of genes that are thought to be necessary but not sufficient for a biologically functional Select Agent genome. For example, we would not worry about the fact that an organism is not merely a bag of genes but contains extensive non-coding regulatory sequence information that would be neglected by a gene-based parts list. Especially in large genomes, we might even choose to simplify a classification system by defining an operationally complete genome as one that has a necessary subset of parts, rather than a complete set. An operational definition of a complete *Bacillus. anthracis*-like Select Agent genome might use a handful of core essential genes (such as ribosomal RNA and DNA polymerase) combined with virulence genes carried on the pXO1 and pXO2 plasmids.

Synthetic genomics raises a complication that someone might substitute some parts of a pathogen for "generic" parts from other genomes—for instance, swapping out genes for core replication functions or swapping out capsid (coat) proteins for different ones while trying to leave alone genes that code for pathogenicity, or moving a toxic gene or pathway into an innocuous non-Select Agent "chassis," such as *E. coli.* We would therefore want a classification system to have a concept of different "resolutions" at which individual parts are defined. For parts that we imagine are more likely to be swappable (such as a DNA polymerase), a part might be "any DNA polymerase," whereas for parts that we think are essential to the Select Agent's virulence, a part might be defined specifically as the Select Agent gene or a closely related modification.

The more modification, the less likely the resulting organism will work as a pathogen (this is related to the prediction problem). It is not at all likely, for example, that DNA polymerases are truly generic and swappable parts (except between closely related organisms), at least not without extensive trial-

[8]It is this difference that makes "classification" feasible and applicable in the context of an oversight system, whereas "prediction" is not. Partial knowledge can usefully be applied for a classification system, while uncertainty and inaccuracy of prediction would result in a system that could not be implemented in a meaningful way.

and-error engineering. But for an operational definition, we can be relatively inclusive and broad about defining a molecular parts list for each Select Agent for classification purposes without worrying too much about whether the assemblage of parts would actually work as a pathogen. We do not have to be concerned about impeding synthetic genomics designs that incorporate parts similar to Select Agents as long as there is no legitimate research purpose for the construct. The regulatory problem is to avoid writing regulations that unnecessarily impede beneficial biomedical research with innocuous organisms and constructs, such as research on vaccines and harmless non-pathogens related to Select Agents or on isolated genes that offer no possibility of resurrecting a complete organism.

It should be scientifically feasible, with current technology and knowledge, to write a clear and acceptable definition of a covered complete Select Agent genome in the form of a genetic parts list, enumerating a subset of key genes (and possibly key regulatory sequences) that are thought to be required for the growth and pathogenicity of the Select Agent. The primary aim would be to define the sequence space around each Select Agent genome including "trivial" modifications that are most likely to encode a similarly hazardous agent without needing highly technical testing and iterative optimization,[9] such as removal of non-essential genes, insertion of foreign genes, or deletion or substitution of non-essential residues. Genes crucial to pathogenesis (in which small variations are known to distinguish Select Agents from non-Select Agents) might be defined more specifically to avoid encompassing vaccines and related non-pathogens, and genes involved in backbone pathways such as replication might be defined more broadly. **Within this framework, it might prove reasonable to define an even broader sequence space to try to account for the most likely synthetic biology scenarios for chimeric agents, such as substitution of core genes for a more generic "chassis" and leaving alone genes for known mechanisms of pathogenesis.**

Any given parts list would reflect only the current state of scientific knowledge about each Select Agent.[10] It would need to be subject to review and revision to keep up with the state of knowledge distinguishing Select Agents from other organisms.

SEQUENCE ANALYSIS OF INDIVIDUAL "PARTS"

Given a parts list that addresses the *content* of a Select Agent genome, the other question is how to define sequences that are covered for each part—the

[9]That is, modifications that would not require work on the scale of an offensive bioweapons research program, which we deem to be beyond the scope of concern for counterbioterrorism in general and for the Select Agent Regulations in particular.

[10]And a parts list would have to be defined for each Select Agent.

genetic *distance* within which a given sequence is defined as being within the sequence space that defines the part, and distinguishes it from related parts of non-Select Agents. There is substantial established methodology regarding this part of the problem.

First we need to explain what commonly used sequence database similarity search programs like BLAST are doing—what their scores mean—before we try to use the scores in an objectively defined classification system. More broadly speaking, what is the state of the art in sequence family classification in computational biology? How are biological sequences assigned to particular sequence groups rather than to others?

BLAST (like related programs) calculates a sequence alignment of the query sequence to a target database sequence, allowing insertions and deletions. It uses a simple scoring system to find a maximum-scoring *local alignment* involving any piece of the query sequence that is aligned to any piece of the target (it does not require the entirety of the two sequences to align over their whole lengths). It calculates two numbers: a total *score*, which is the sum of all the residue alignment scores and gap penalties in the optimal-scoring local alignment, and an *E value* (expectation value; or *P value*, probability value, which is essentially the same sort of number) which represents the statistical significance of the total score.

A BLAST score constitutes an answer to the specific question: Is the target sequence homologous (related by evolutionary common ancestry) to the query sequence? The higher the score, the more statistical evidence that supports an inference that the query is homologous to the target.

The E value tells us how many times we should have expected to see a score that high if we searched a *random* target sequence database of the same size but consisting only of nonhomologous (unrelated) sequences. If the E value is low (say, 0.01 or less), we say that the hit is statistically significant; that is, that it was unlikely to have occurred by chance.

Neither a BLAST score nor an E value is a "distance" suitable for sequence classification into families. A BLAST score measures how likely it is that two sequences are related at all, not how closely related they are, and the E value just measures the statistical significance of the score. For example, the longer the alignment is, the more statistical signal in favor of homology accumulates, the higher the score, and the lower the E value. It is easier to find high-scoring, more significant alignments for long proteins than for short ones. However, a measure of genetic distance between two sequences should be independent of the lengths of the aligned sequences.

The standard unit of distance between two sequences in phylogenetic (evolutionary) inference is *residue substitutions per site*. Assuming that two sequences are homologous and assuming that we have an alignment of each individual homologous pair of residues in the two sequences, we infer how many residue substitutions have occurred at each aligned residue in the evolutionary time since the divergence of the two sequences. At short distances, simple percent identity of the alignment approximates the evolutionary distance; at longer distances, more sophisticated models are used to convert the observed dissimilarity between two sequences into a distance estimate. Thus perhaps counterintuitively, simple percent identity (not percent "similarity," not BLAST score, and not E value) is a reasonable although rough measure of genetic distance. A caveat in using simple percent identity of a BLAST alignment as a genetic distance is that BLAST local alignments tend to identify and align different regions of different targets to the same query, and some regions of a protein or DNA are more conserved (and therefore more highly identical) than other regions.

Different sequences evolve at different rates. (Indeed, even different residues in the same sequence family will evolve at different rates; the residues in a critical enzymatic active site may be highly conserved, and residues in a solvent-exposed surface loop of a protein may be highly variable.) That is why it is not possible to define an overall threshold for genetic distance (or percent identity) that separates one sequence family from another (or one species from another) that works for all sequence families or all species.[11]

For evolved sequences, the distances can be used to infer a phylogenetic tree. A phylogenetic tree is the best representation of the hierarchy of relationships between a set of sequences. There is no natural discrete classification of a set of homologous sequences into a subset of families. Breaking the set of sequences into more than one discrete family and assigning some sort of (taxonomy-like) name to these subsets must always be arbitrary, relying on some operational goal. Different reasonable people might break the same tree into different subsets to serve different goals.

Over evolutionary time, biological function changes in ways that are not fully predictable from evolutionary distance. A single residue change can change an agonist, an activator of some function, to an antagonist that blocks that very function. Exactly the same protein sequence can be found serving completely different functional roles, such as central metabolic enzymes that also function as "crystallins" to build the transparent lens in an eye, or iron-requiring metabolic enzymes that also serve as regulatory proteins that turn other genes on and off in response to changing iron concentrations. It is only *usually* the case that homologous proteins have similar functions; the more closely related they are, the more likely they are to have more similar functions.

Those issues bring us back to the crucial difference between prediction and classification. We can *classify* a new sequence as being within a group of known sequences on a tree; we cannot necessarily *predict* the function that the

[11]That is, just as content (or a parts list) must be defined for each Select Agent, the distance must also be determined for each part of each Select Agent.

sequence encodes. However, as long as our classification scheme does not mislabel a non-Select Agent as a Select Agent, it may suffice operationally.

METHODS FOR SEQUENCE SUBFAMILY CLASSIFICATION

Armed with background about such sequence comparison programs as BLAST, evolutionary distances, evolutionary trees, and functional sequence subgroups as distinct clades on trees, we return to the problem of screening DNA sequence orders for Select Agents.

Screening for significant BLAST hits to a database of sequences of concern does not work, because the parts of Select Agents are homologous to parts of many non-Select Agent organisms. Many non-Select Agent organisms would trigger false-positive results.

No single threshold on BLAST score or E value solves this problem, because (to give just one reason) scores and E values depend on the length of the alignment, but different sequences of different genes have different lengths. No single threshold of percent identity will work, because some proteins are less conserved in sequence than others (that is, more tolerant of sequence differences while retaining function).

Intuitively, we would like to identify when a sequence is "closer" to a Select Agent sequence than to a non-Select Agent sequence. To do that, we must have a database not only of Select Agent sequences of concern, but also of at least a representative set of homologous non-Select Agent sequences that we are trying to distinguish from Select Agent sequences.

The simplest thing to do with the combined database is to look at the best BLAST hit for a new sequence: If the best BLAST hit is a Select Agent sequence, the new sequence is "closer" to a Select Agent than not.[12]

"Best BLAST match" classification is a reasonable and often used strategy, but it does not represent the state of the art in inferring subfamily classifications. Most important with "best match" it is hard to distinguish the case in which the new sequence belongs to a known subfamily (say, a Select Agent gene) from the case in which the new sequence is a member of a new subfamily (say, a homologous non-Select Agent gene) that is not represented in the database yet. For example, if a sequence is only 30 percent identical to a Select Agent coding gene and 20 percent identical to the nearest known non-Select Agent homolog, but all known variants of the Select Agent gene are greater than 95 percent identical to each other, it is likely that the sequence is from a previously undescribed organism (because it falls well outside the expected range of variation within the known Select Agent sequences). A simple "best match" criterion nonetheless classifies the sequence as a Select Agent. That could be problematic if newly

[12]This is the method outlined in the DHHS Screening guidance for DNA Synthesis Companies (DHHS, 2009), and the combined database that it suggests is NCBI-Entrez.

discovered organisms came to be classified as Select Agents solely because their closest known homolog in the database was a Select Agent, regardless of genetic distance or the disease-causing phenotype of the organism.

The state of the art in subfamily classification is called phylogenomic analysis. The position of the new sequence in a tree of known homologous sequences that represent different functional subfamilies (or Select Agent and non-Select Agent sequences) is examined. If the new sequence falls within a clade on the tree that defines the observed range of variation in a subfamily, one can assign it to that subfamily with confidence. If the new sequence falls outside any known subfamily but is still homologous with the overall family, one would not assign it to a subfamily and instead would annotate it according to the more generally conserved features of the larger family. Phylogenomic analysis formalizes the intuitive idea of classifying sequences with their closest sequence relatives, while being careful to see the case in which a sequence is "novel," falling outside the known range of variation of known nomenclature groups.

Phylogenomic analysis is usually done manually by human experts. The process of making high-quality phylogenetic trees involves several steps (collecting representative sequences, making an accurate multiple alignment, and inferring a reliable sequence tree) that benefit from human judgment. There is a good deal of research into automating phylogenomic analysis, but one would probably not now seek to develop an automated high-reliability classification system based on explicitly tree-based phylogenomic inference.

The state of the art in *automated* subfamily classification uses models called profiles to represent each different subfamily. Especially in probabilistic form as what are called profile hidden Markov models (profile HMMs), a profile is a model of the space of possible sequences that belong to some group. Representatives of the sequence group are identified and aligned into a multiple sequence alignment. Given a multiple sequence alignment, such software tools as HMMER build a profile of the alignmentin which each aligned column is represented by a position-specific scoring system that captures information about how conserved each residue is in any new homologous sequence that would belong to the sequence family. Profile HMMs calculate the likelihood that a new sequence "belongs" to the profile's family. A comparison of two profile HMM scores is a hypothesis test for whether a new sequence belongs to one profile or the other. It allows experts to define protein or DNA sequence families of interest and to prebuild profiles that represent the sequences that belong to those families. New sequences are compared with profile libraries and automatically classified on the basis of profile score.

Profile-based classification is related to phylogenomic analysis. The main difference is that the phylogenetic tree is taken into account by human experts in building an appropriate set of profiles *before* any new sequences are analyzed; afterwards, the profile itself does not use an explicit phylogenetic tree. An expert identifies which subfamilies are of interest for classification and builds a profile for each subfamily and for the entire family (and possibly at different levels of the hierarchy of the phylogenetic tree, capturing progressively smaller and more detailed levels of subfamily annotation).

For example, if a sequence analyst wanted to distinguish NAD-dependent malate dehydrogenases, NAD-dependent lactate dehydrogenases, and novel NAD-dependent dehydrogenases that might work on a novel substrate, she might produce a profile of known malate dehydrogenases, a second profile of known lactate dehydrogenases, and a third profile of all known NAD-dependent dehydrogenases. If a new sequence is an NAD-dependent dehydrogenase, it will probably get a significant score against all three profiles (because malate and lactate dehydrogenases are evolutionarily homologous and structurally similar). If it is a lactate dehydrogenase, the highest score will (should) be to the lactate dehydrogenase profile; if it is a novel dehydrogenase, the highest score should be to the general dehydrogenase profile.

Large profile databases are in widespread use for protein-gene annotation. A good example of a large collection used for microbial gene function annotation is the TIGRfams database at the J. Craig Venter Institute. For example, in TIGRfams, NAD-dependent L-lactate dehydrogenases are represented by model TIGR01771 (L-LDH-NAD), and bacterial NAD-dependent malate dehydrogenases are represented by model TIGR01763 (MalateDH_bact). TIGRfams does not include a general dehydrogenase profile to catch cases of novel dehydrogenases; instead it uses a strategy of predefining a curated score threshold for each family.

Different profile databases are in widespread use, aiming at different annotation goals with different levels of protein classification. TIGRfams is an example of a profile library that aims at functional protein subfamily classification. In contrast, databases like Pfam and SMART aim at the "superfamily" level, aiming to capture the widest possible diversity of all remote homologs of protein families, regardless of functional classification. Pfam, for example, has a single profile (PF02615, Ldh_2) that classifies all NAD-dependent dehydrogenases into a superfamily that includes both L-lactate and malate dehydrogenases. Almost all profile libraries are based on the same underlying profile HMM approaches and use one of two software packages (SAM or HMMER).

Profile-based sequence classification is highly automatable because the classification system relies on a relatively stable set of sequence alignments of representative sequences that define the desired families. Thus, the classification system is relatively robust to the exponential growth of the sequence databases. In contrast, classification systems that rely on all-versus-all comparison of all known sequences tend to be unstable, and it is difficult to ensure their quality because the sequence database changes rapidly. A profile-based system can be developed, tested, and benchmarked with care by experts and then maintained and used stably over long time periods.

OUTLINE OF A POSSIBLE SYSTEM FOR PROFILE-BASED CLASSIFICATION OF SELECT AGENTS

With current sequence analysis technology, it would be possible to develop an automated and precisely defined system for classifying genome sequences as to whether they are "complete" Select Agents. The system might look something like the following.

For each Select Agent, a *minimal parts list* would be identified by experts. It would be the set of genes that are thought necessary to make an infectious Select Agent genome. The parts list does not have to be exhaustively complete, because it is being used only to classify genomes, not to describe them fully; we might choose to include only a representative subset of the genes in a microbial genome to reduce the size of the classification system (and the work needed to create it). A genome that contained all the parts on the minimal parts list would be classified as a "complete, infectious" genome for operational purposes of the Select Agent Regulations. A genome that did not contain all of the parts would be a "genomic fragment" for the purposes of the Select Agent Regulations.

For each part, an automated *profile-based classification system* would be developed to differentiate the subfamily of sequences belonging to the Select Agent from the larger family of sequences belonging to non-Select Agent organisms. The specificity of these profiles would vary, some of them being very specific to only the Select Agent, and some of them being general and encompassing both Select Agent and non-Select Agent sequences. This step requires expert judgment. The more general models would allow the classification system to deal with the possibility that some parts are substitutable (by synthetic biologists) with "generic" parts, so these profiles might be made at a more generic level—any RNA-dependent RNA replicase, rather than specifically the Select Agent RNA-dependent RNA replicase, for example. The more specific models would focus on the parts thought to be most responsible for pathogenicity as opposed to core replication, metabolism, and growth functions. **These specific models (as distinguished from the generic models) might be specially flagged to raise a flag to indicate that parts of a Select Agent are present even though a complete Select Agent genome is not**, for the purposes of prudent follow-up on the part of a DNA synthesis company—for instance, if an order might represent an attempt to obtain a Select Agent genome in several individually legal pieces.

For each Select Agent, given a minimal parts list and a profile-based classification system for each part the classification system would be tested, benchmarked, and challenged using known genome sequences. To be useful, the classification system would be required to classify correctly all known sequence variants of a Select Agent (and a set of reasonably imaginable ones), and a representative set of the most closely related non-Select Agent genomes, including very close relatives, such as vaccine strains or non-pathogenic variants used in laboratory research.

For good classification, it is not sufficient to know a single representative genome sequence of each Select Agent. Using a classification system is an attempt to determine whether a novel sequence fits into the "cloud" of sequences representing expected genetic variation for a Select Agent genome, as opposed to the "clouds" of sequences representing the most closely related non-Select Agents. The more sequences are known, the better the expected genetic variation will be understood. Genome sequences of almost all Select Agents are available, but there has been less emphasis on obtaining genome sequences for closely related non-Select Agents. Future studies are sure to discover numerous new microbial and viral species, and it is desirable that these new discoveries not be misclassified as Select Agents just because they are closely related to Select Agents. **More systematic genome sequencing of non-Select Agents would improve our knowledge of biodiversity and would be useful in developing a good classification system.**

The profile classification system would have to be reviewed and revised, as new knowledge accrued that required newly discovered Select Agent or non-Select Agent variants to be classified. The updating process would resemble the continuing curation of other profile library classification systems, such as TIGRfams and Pfam.

Because it is automatic and software-based, the classification system could be made readily and transparently available on the Web, where it could be reviewed and challenged by scientists in the community to be sure, for example, that it was not inadvertently misclassifying useful non-Select Agents, such as vaccines and attenuated research strains.[13]

Timely testing, updating, and public review of the system would guard against classification errors. Automated annotation of protein function based on sequence similarity analysis is robust but not error-free (Schnoes et al. 2009).

This essentially phylogenetics-based system will work better for some Select Agents than for others. The greatest difficulty in clasifying Select Agents with a phylogenetic subfamily system will occur in cases in which very closely related viruses in the same phylogenetic group that have small, easily evolved genetic changes that differentiate highly pathogenic Select Agent strains from low-pathogenicity non-Select Agent strains, that come and go in a phylogenetic tree; the "high-pathogenicity" avian influenza viruses are an example. Similar cases of convergent functional evolution arise in protein function annotation, in wihch a small number of changes in active site residues can shift a protein function and these changes convergently evolve multiple times in multiple lineages. Alternative methods that key on critical functional residues have been

[13]This would require a process for reclassifying an agent in response to input from the scientific community.

developed to deal with the problem for protein function annotation (Hannenhalli and Russell 2000) and could be deployed and benchmarked for the Select Agent classification problem.

There will be cases in which any sequence-based classification system must fail altogether. For example, the bovine spongiform encephalopathy (BSE) prion agent is on the Select Agent list, but prions are an alternatively folded conformation of a host protein; the amino sequence of the prion form of the protein is identical to the benign host form. The only way to distinguish the BSE prion from the natural host protein is by experimental assay.

There would be no pretense of prediction in this classification system. Many genomes would be classified as a Select Agent because they have all the parts of a Select Agent, but there is little reason to think that all those parts would necessarily work in concert to produce a working, infectious, pathogenic organism; indeed, most synthetic genomes that had all the independent parts would probably *not* work as dependent wholes. From the standpoint of dealing with the implications of synthetic biology and synthetic genomes, the utility of the classification system would not be to distinguish successful genome designs from unsuccessful ones—"bootable" pathogens from inert DNA sequences—but to distinguish attempts to synthesize a dangerous genomes similar to a Select Agent from an attempt to synthesize benign genomes from a non-Select Agent organism, a non-pathogenic strain, or a vaccine. **The classification system does not distinguish legitimate research from illegitimate research; rather it identifies agents that are restricted under the Select Agent Regulations and provides a means of identifying "sequences of concern" that may be worth monitoring.**

The goal of the Select Agent Regulations is to restrict availability of the most dangerous *known* pathogens while not impeding beneficial biomedical research on known or emerging pathogens. In dealing with synthetic biology and the potential threat posed by novel agents, our goal is to try to regulate the most obvious attempts to synthesize a potentially working pathogen, and the current state of the art in synthetic biology is the ability to produce new combinations of existing biological parts, not to devise new genomes entirely *de novo*. We can never exclude radically novel synthetic biology designs, but we can raise the bar to the point where bioterrorists would have to possess knowledge *better* than the current state of the art with respect to what biological parts are necessary in a pathogen to evade a parts-list-based Select Agent classification system or would have to engage in an offensive biological weapons research program on a scale that would come under the Biological Weapons Convention.

A classification system would clearly be the easiest to develop for Select Agents with the smallest parts lists. The easiest would be the protein toxins composed of one or a few proteins, such as abrin and ricin. The next easiest would be the proteins encoding the multistep synthetic pathways for metabolite toxins such as diacetoxyscirpenol, saxitoxin, and tetrodotoxin (on the presumption that this biosynthetic pathway might be moved in a modular form into a new host to create a new organism that expresses the toxin). Next would be the viruses, ranging from small genomes (such as Lassa virus) to large ones (such as smallpox). The microbial genomes would be the hardest to deal with, and would require the most thought about what parts are generic and what parts are specific to a Select Agent pathogen.

CONSIDERATIONS FOR IMPLEMENTATION OF A PROFILE-BASED CLASSIFICATION SYSTEM

It is not the role of our committee to recommend specific implementation plans, nor are we properly constituted to do so. But we were tasked with describing an "alternative framework" for oversight, so it is appropriate to make some observations about implementing a profile-based sequence classification system along the lines discussed in this chapter.

To be useful for unambiguous regulations, there would need to be a single agreed-on classification system as opposed to multiple competing systems developed by different research groups. That would require a centralized funding plan that would balance the benefits of single source standardization by a single Select Agent classification system team against the need for oversight and review to maintain quality and efficiency in the absence of peer competition.

A classification system would require a small team of full-time staff to develop and maintain it. The sequence curation work required is substantial. Classifying the current 82 Select Agents would require 82 parts lists and on the order of several thousand different profiles for the parts, and each Select Agent classification would need to be carefully tested and maintained over time. That would be on the same scale as the curation effort involved in the current Pfam or TIGRfams databases for automated protein sequence annotation. The Pfam database, for example, consists of about 12,000 profiles of common protein domain families, maintained by four to six skilled full-time staff since the mid-1990s, including sequence analysts, database administrators, and software developers.

The curation team would need advice from a panel of leading scientists for each group of pathogens. The scientific advisory panels would need to meet regularly to review the relevant literature and research results and would need to develop and maintain up-to-date consensus on the parts lists and parts classifications that define suitable sequence spaces around each Select Agent. These defined sequence spaces would be embodied in an automated classification system by the curation team. The classification system (and comments from the scientific community on its accuracy, gathered from the scientific community) could be reviewed by the appropriate government departments, and the database system would be approved as guidance in interpreting the law, much as the Centers for Disease Control issued a written document to guide gene synthesis companies in interpreting the application of the Select Agent Regulations to synthetic DNA.

These scientific advisory panels would probably include not only U.S. scientists, but the best scientists from around the world. International participation would have intangible additional benefits. Gene synthesis is an international industry; international harmonization of regulation and best practices for biosafety and biosecurity in synthetic genomics is an important area. In addition, participation of international scientists in the undertaking could raise awareness of dual-use issues among international researchers—a major objective of the National Strategy for Countering Biological Threats[14] and of the NSABB.

A balance would need to be struck between the need to keep definitions up to date with the state of scientific knowledge about the genetic composition of plausible complete and infectious Select Agent pathogens and the need to have a stable regulatory environment. It would be undesirable to have high-consequence regulations like the Select Agent Regulations changing on a rapid time scale. It would be unreasonable, for example, to have sequences moving on and off the list on a time scale much faster than the time scale of converting a laboratory to meet the Select Agent Regulations. A suitable time scale might be to issue an updated classification system every two years. This is consistent with the current review process for the Select Agent Program, which is overseen by the Intragovernmental Select Agents and Toxins Technical Advisory Committee (ISATTAC).

The periodic expert review and update cycle could be meshed well with recommendations of other recent advisory reports calling for increased cross-agency harmonization of the Select Agent Regulations, and for increased transparency in the procedures for moving agents on and off the list.[15]

As we have discussed, the decisions made in establishing classification boundaries in sequence space are unavoidably arbitrary. They cannot be interpreted as biological *predictions* of whether given synthetic genome sequences would function as dangerous pathogens. Nevertheless, such a system would be an improvement over the current process. It would transparently, consistently,

[14]"Transform the international dialogue on biological threats: Activities targeted to promote a robust and sustained discussion among all nations as to the evolving biological threat and identify mutually agreed steps to counter it."

[15]2009 National Research Council report *Responsible Research with Biological Select Agents and Toxins*, "RECOMMENDATION 2: To provide continued engagement of stakeholders in oversight of the Select Agent Program, a Biological Select Agents and Toxins Advisory Committee (BSATAC) should be established. The members, who should be drawn from academic/research institutions and the private sector, should include microbiologists and other infectious disease researchers (including Select Agent researchers), directors of BSAT laboratories, and those with experience in biosecurity, animal care and use, compliance, biosafety, and operations. Representatives from the federal agencies with a responsibility for funding, conducting, or overseeing Select Agent research would serve in an ex officio capacity . . ." (NRC 2009b).

and unambiguously represent the harmonized views of a community of experts. A centralized system would almost certainly lead to better decisions than reliance on a series of dispersed judgments by individual scientists in gene synthesis companies who have little specific knowledge about the pathogen sequences that they might be asked to synthesize.

It would be prudent to develop the system in phases, starting with a pilot project on a subset of the smallest Select Agent sequences, such as the protein toxins and the smallest Select Agent viruses. Several of the smallest Select Agent sequences are at the same time considered the most dangerous and feasible threats for current synthetic genomic technology, and also the smallest and easiest test cases for a profile-based classification system (because they require definition of the fewest parts).

ROLE OF PREDICTION AND CLASSIFICATION IN BIOSAFETY

Chapter 2 discussed how high-level biological phenotypes, such as pathogenicity and transmissibility, cannot plausibly be predicted with the degree of certainty required for statutory biosecurity regulations, either now or in the foreseeable future. Nonetheless, predictive ability is a major goal of biology, and it is sure to develop. Any predictive ability will develop first in ways suited for probabilistic risk assessment—for raising potential biosafety concerns about an encoded organism—long before prediction reaches the far higher bar of making completely precise and accurate judgments suitable for declaring that novel synthetic constructs are dangerous weapons and immediately subject to Select Agent Regulations in the absence of actual experience with the encoded organism's properties. Predictive systems biology will slowly give us a better ability to assess whether a synthetic genome sequence *might* be more or less hazardous and whether the organism that it encodes *might* be more or less likely to have the phenotypic properties of Select Agents. Therefore, to the extent that prediction of biological properties of Select Agents does become possible, we believe that it will first be useful in the context of a yellow flag biosafety warning system—warning investigators and their institutions and institutional biosafety committees (IBCs) that a synthetic construct might be hazardous, perhaps inadvertently so, and that the construct and its encoded organism ought to be handled with an appropriate level of biosafety containment.

The centralized classification system for pathogen sequences described above could inform biosafety judgments long before prediction has a substantial impact. The two concepts of a minimal parts list for a complete pathogen and a profile-based classification system could help scientists to identify constructs that, although not complete pathogens within the definitions of the Select Agent Regulations, merit close scrutiny for their biosafety and biosecurity implications. Classification is not prediction, but it is completely plausible that constructs that are close to a minimal parts list for a pathogen or are similar to a pathogen gene merit more thorough biosafety evaluation than constructs that are not.

RAISING A YELLOW FLAG FOR "SEQUENCES OF CONCERN"

A wide array of DNA sequences are in a category that, although not completely innocuous, should not be subject to the severe constraints on research imposed by the Select Agent Regulations. The Select Agent Regulations cover complete genomes ("nucleic acids that can produce *infectious forms* of any of the Select Agent *viruses*"—language that has been formally interpreted as excluding "genomic fragments from Select Agents"). There are many reasons for prudent concern about providing synthetic genomic fragments of Select Agents. A "bad actor" might seek to obtain a genome in fragments that individually did not meet the criteria of the Select Agent Regulations and then assemble the complete genome. The "bad actor" would be in violation of the Select Agent Regulations by constructing a complete genome, but from a regulatory perspective, we might want to make this technically plausible scenario more difficult to carry out and might want to have more advance notice when such a scenario might be playing out.

We would like to explore the concept of a "yellow flag" system that could provide a transition between the standard biosafety practices that are applied to the vast majority of research projects and the highly regulated, highly restrictive practices required by the Select Agent Regulations. (The concept of the yellow flag will also be discussed in Chapter 4 and Appendix L.) Our best chance to reduce public health risks posed by infectious disease is an active, efficient, and safe R&D program directed at known and emerging pathogens. **This committee and others believe strongly that overbroad application of the Select Agent Regulations will increase risks to public health by increasing the cost and reducing the speed of critically important research on pathogens. The aim of the yellow flag is to provide a set of practices that improve safety and security without increasing the number of sequences that are covered by the Select Agent Regulations.**

Restrictions on pathogen research have two primary goals: to make it harder for bad actors to use pathogens as weapons or as tools for bioterror and to avoid the accidental, inadvertent, or ill-advised production of hazardous constructs by well-meaning investigators. Meeting both goals through a yellow flag system probably requires some form of disclosure of the pathogenic sequences or experimental plans to scientists outside the groups carrying out the research. Thiat disclosure could involve centralized reporting of yellow flag sequences, public disclosure, or some kind of institutional review. A centralized system for reporting yellow flag sequences would allow detection of the simplest scheme for avoiding the Select Agent Regulations—splitting the order for a viral genome or a toxin between two different gene synthesis companies. It would also enable intelligence analysis and monitoring of DNA sequence orders. Public disclosure would offer the virtues of an open-source system—the power of review by many different observers—but it would require a substantial change in the culture of science. A standardized system for identifying "yellow flag" sequences coupled with biosafety review (and biosecurity monitoring) might offer the simplest approach to reducing risks.

We envision that the actions taken in response to a yellow flag could be informal, prudent best practices in that they fall outside the strict regulatory boundaries of the Select Agent Regulations. It would also be possible to use a yellow flag system in more formal ways. For example, an IBC or funding sponsor could ask that yellow-flagged synthetic constructs trigger some sort of special notification for purposes of oversight if for no other reason than to track what laboratories were in possession of such constructs. Similarly, DNA synthesis companies might be asked to maintain records of yellow-flagged constructs that they provide, to facilitate forensic investigation in the event of the criminal construction of a complete Select Agent from synthetic parts.

The concept of raising a yellow flag on synthetic genomic constructs clearly overlaps with the biosecurity goals of the Select Agents Program—providing a sort of buffer zone that identifies individual Select Agent synthetic parts that do not rise to the precise inclusive definitions of the "complete, infectious" genomes that come under the Select Agent Regulations. Moreover, we view a yellow flag system more broadly from a biosafety perspective. We believe on intuitive grounds that the probability of accidental, inadvertent, or ill-advised hazardous constructs by well-meaning investigators (including hobbyist biohackers) is much higher than the probability of deliberate weapons constructs,[16] although neither probability can be estimated. A yellow flag system could help to build a web of reasonable oversight of synthetic genomic constructs. Investigators or biohackers could be warned if their constructs included potentially dangerous Select Agent-like parts, to be sure that they were aware and to be sure that they were prepared and equipped to handle the constructs appropriately. It would also be possible to expand the list of yellow flag parts used in profile-based classification beyond the Select Agents list to include other sequences of concern.

Initially, the "yellow flag" system could be based on the profile-based classification system described above, and best practices could be promulgated for how investigators, institutions, IBCs, and synthetic genomics companies should deal with constructs deemed to be "potentially risky." As predictive methods are developed, they would naturally augment such risk assessments. Because the profile-based classification system would be transparently available in its func-

---

[16]At the very least, those engaged in a deliberate weapons program are less likely to use DNA sysnthesis companies as a means of obtaining genomic fragments, when those companies are known to screen orders for "sequences of concern."

tion for the Select Agent Regulations, it would also be transparently available for parallel development of biosafety best practices of the yellow flag variety. That is, community awareness and best practices for responsibility, review, and oversight of biosafety of synthetic constructs could be developed now, starting with the profile-based classification system, and this biosafety framework would then be in place as scientifically sound *de novo* predictive systems develop.

SHOULD SUCH A SYSTEM BE BUILT?

All that said, *should* such a profile-based classification system be built? Our committee is not constituted to answer that question. Our answer to this part of our charge is that it *could* be built.

Together, the profile-based classification for Select Agents, and the yellow flag system for sequences of concern would address many of the emerging biosecurity concerns posed by synthetic biology. Gene synthesis companies, which have to make daily judgments based on the Select Agent Regulations and other regulations, strongly favor development of such a system. As discussed in Chapter 1, several of the companies have agreed to a common set of procedures for screening customers and sequences. In addition to helping to identify sequences that qualify as Select Agents, the parts list profile-based classification system would provide information to be used for flagging sequences of concern. The companies have agreed to apply enhanced customer screening to orders that do not meet the definition of the Select Agent Regulations but do include Select Agent sequences or other pathogen sequences. The goal of such screening is to supply sequences only to customers that have a legitimate research use for them and have the resources to handle them safely.

The system could also have utility that is independent of its role in clarifying the regulation of pathogen sequences. **One of the most consistent lessons of modern biological research is that cross-pollination between diverse fields of expertise consistently yields valuable scientific insights and useful new tools (NRC 2009b). Gathering information on the most important human and animal pathogens into a single, consistently annotated database will make it easier for scientists and engineers who are not experts on a particular pathogen to develop new diagnostics, vaccines, and therapeutics.**

However, the system we have described does have some objectionable qualities. It may be "overkill." It erects a moderately complicated and pedantic definitional framework around the purely regulatory question of what sequences represent complete, infectious Select Agents or not.

One might take an alternative view that the regulatory language could be clarified more simply and elegantly with a language of "reasonableness," expressing the concepts of sequence classification methods we have described here but without the potential heavy-handedness of a centralized automated computational implementation. For instance, there might be regulations and guidelines that declare the intuitive notion that a synthetic genome is "reasonably" expected to be complete and infectious and that the sequence of its parts would appear to a "reasonable" person to be closer to that of a Select Agent reference sequence than to that of the nearest non-Select Agent reference sequence. Such language alone would be an improvement over the use of problematic percent identity thresholds in some current regulatory language, and over the absence of any guidance at all for dealing with modified or chimeric Select Agent synthetic genomes. The intuitive concepts of sequence-based classification are sufficiently clear for anyone to know whether a sequence is in the vicinity of any reasonable definition of the line. However, a "reasonableness" approach does not solve the problem of vagueness that troubles the DNA synthesis companies, researchers, and law enforcement as they try to apply the Select Agent Regulations. Without a precise definition of the sequences covered by the Select Agent Regulations, companies might choose to reject any construct that contains Select Agent-related sequences, and researchers are left unsure whether they are subject to the Select Agent Regulations. Moreover, the reasonableness approach does not begin to address the issue of novel synthetic constructs and sequences of concern, which are a motivator for the development of an alternative framework for oversight.

4

Committee Findings and Conclusions

This chapter summarizes key findings and major conclusions. As discussed below, the committee finds that it is not feasible to develop an accurate oversight system based on prediction. However, a gene sequence based classification system for Select Agents and a "yellow flag" biosafety system for "sequences of concern" could be developed with current technologies. The classification system discussed in Chapter 3 (see also Appendix L) could provide much needed clarification regarding application of the Select Agent Regulations. The "yellow flag" system could provide a means of guidance and oversight for "sequences of concern." The "yellow flag" system would function as an extension of biosafety; however, because it is not regulatory in nature, it could provide information relevant to biosecurity in a more dynamic and timely fashion than the Select Agent Regulations.

The committee has identified crucial components that would enable such systems. Although the individual near-term milestones, as described, may be beneficial to scientific progress and would probably improve the current biosafety and biosecurity system, careful consideration should also be given to the limitations and challenges of developing and implementing these or similar systems.

FINDINGS AND CONCLUSIONS

(1.) ***Purpose of the Select Agent Program: The Select Agent Program is intended to restrict access to known agents that pose a threat to* biosecurity.**

(a.) **The Select Agent Program is intended to focus on biosecurity, rather**

than biosafety.[1] As discussed in Chapter 1, biosafety and biosecurity are related and complementary, but there are important distinctions. *Biosafety in Microbiological and Biomedical Laboratories* (CDC/NIH 2007) defines biosafety programs as those which "reduce or eliminate exposure of individuals and the environment to potentially hazardous biological agents," whereas the objective of biosecurity is to "prevent loss, theft or misuse of microorganisms, biological materials, and research-related information" (CDC/NIH 2007). Biosafety is reducing the risk that pathogens or toxins will escape containment and cause illness in researchers, clinicians, or the general public. Biosecurity is related to minimizing the possibility that such pathogens will be used for malevolent purposes.[2] The BMBL sets standards for how U.S. laboratories conduct research with biological agents and toxins; the Recombinant DNA Advisory Committee (RAC) and Institutional Biosafety Committees (IBCs) provide guidance and oversight focused on biosafety.[3] This is a robust and effective system. In fact, if the purpose of the Select Agent Regulations were solely biosafety, it would largely be an unnecessary duplication. However, the primary role of the Select Agent Regulations is to restrict access to agents that may be used as biological weapons by someone with nefarious intent. There is a good deal of overlap between biosafety and biosecurity threats in that the pathogens deemed to pose

[1]This section draws on discussion in the 2009 National Research Council report "Responsible Research with Biological Select Agents and Toxins."

[2]The 2009 National Research Council report states:"[i]t should be noted that the use of the term "biosecurity" presents a number of difficulties. At its most basic, the term does not exist in some languages, or is identical with "biosafety"; French, German, Russian, and Chinese are all examples of this immediate practical problem. Even more serious, the term is already used to refer to several other major international issues. For example, to many "biosecurity" refers to the obligations undertaken by states adhering to the Convention on Biodiversity and particularly the Cartagena Protocol on Biosafety, which is intended to protect biological diversity from the potential risks posed by living modified organisms resulting from modern biotechnology. (Further information on the Convention may be found at <http://www.cbd.int/convention/> and on the Protocol at <http://www.cbd.int/biosafety/>.) "Biosecurity" has also been narrowly applied to efforts to increase the security of dangerous pathogens, either in the laboratory or in dedicated collections; guidelines from both the World Health Organization (WHO 2004) and the Organization for Economic Cooperation and Development (OECD 2007) use this more restricted meaning of the term. In an agricultural context, the term refers to efforts to exclude the introduction of plant or animal pathogens. (See NRC 2009a:8-9 for a discussion of this and other issues related to terminology.) Earlier NRC reports (2004ab, 2006, 2009a) confine the use of "biosecurity" to policies and practices to reduce the risk that the knowledge, tools, and techniques resulting from research would be used for malevolent purposes."

[3]The Recombinant DNA Advisory Committee (RAC) assisted the NIH in the development of the NIH Guidelines for Research Involving Recombinant DNA Molecules, which has become the standard of safe scientific practice in the use of recombinant DNA. Institutional Biosafety Committees (IBCs) which are mandated by the NIH Guidelines, are charged with reviewing research involving recombinant DNA, although many IBCs have chosen to review other forms of research that involve potential biohazards—including research involving Biological Select Agent and Toxins (BSATs). Institutions are required to register their IBCs with NIH's Office of Biotechnology Activities.

as the greatest biosafety risk (BSL-4 agents) are all Select Agents; however, not all Select Agents pose substantial risk to individual public health (for example, BSL-2 agents may be Select Agents). An agent may pose a security risk because of its potential for weaponization or adverse economic consequences, rather than direct effect on human health.

Handling of Select Agents requires controlled access to facilities, physical security, inventory control, and site-specific risk assessments. Everyone who has access to Select Agents must be cleared through the Federal Bureau of Investigation's Criminal Justice Information Services Division with a background check. Failure to meet the requirements may result in criminal penalties of fines and up to 10 years of imprisonment. Thus, the Select Agent Regulations can be reasonably viewed as an instrument of law enforcement to facilitate attribution[4] and prosecution in the event of domestic use or, deliberate or inadvertent possession of potential biological weapons.

(b.) **The Select Agent Program necessarily focuses on the known.** The Select Agent Program is intended to limit access to agents that there is reason to believe could be used as weapons—essentially to remove the "low-hanging fruit" and make it more difficult for persons with nefarious intent to obtain or create a bioweapon. The Select Agent Regulations work primarily and most effectively in the context of possession and transfer of known stocks—providing a "chain of custody" [5]—in which names and Select Agent status are propagated

[4]Microbial forensics plays an important role in attribution efforts. Microbial forensics, also called bioforensics, is a relatively new scientific discipline that draws from other disciplines including genomics, microbiology and plant pathology. Microbial forensics is dedicated to analyzing microbial activity as evidence for attribution purposes and backtracking. Microbial forensics procedures support 'decision taking' at biosecurity levels, follows strict chain of custody of specimens and demands a rigorous (accredited) and unbiased performance. Therefore, microbial forensics includes the complete range of forensic evidence analysis from microorganisms to associated evidence materials found at the site of a suspected outbreak or crime scene.

[5]The term *chain of custody* can be used on several different scales of resolution. On the grossest scale, it is sometimes used to describe the provenance of a physical sample of an organism. For example, after the 2001 anthrax letter attacks, a large effort was expended to attempt to trace the provenance of all the Ames strains at laboratories in the United States. Lack of comprehensive recordkeeping before the Select Agent rules made that a difficult and imprecise process at best. On a medium scale, *chain of custody* as applied to a single laboratory now is used to mean the strict recordkeeping that allows knowledge of which staff had access to the locked laboratories and locked freezers or cabinets where the Select Agent organisms are stored and knowledge of all dispositions of materials that have taken place within the laboratory. On the finest scale of resolution, *chain of custody* has a specific meaning related to evidence handling for a potential court case: written records of each person and each procedure followed and sealed evidence bags documenting all the physical containers that held the Select Agent material in the different process steps (tubes, plates, and so on). In this report we refer mainly to the two larger-scale resolutions of chain of custody.

in a well-defined manner from registered sender to registered recipient of Select Agent cultures.

As discussed throughout this report, Select Agents are defined according to a taxonomic list of known bacteria, viruses, toxins, and fungi. Novel agents, whether natural or synthetic, are not covered by the Select Agent Regulations. When a novel agent emerges, it is named, and research is initiated to study its mechanism of action, the potential threat that it presents, and its susceptibility to countermeasures. After knowledge is obtained, an agent may be considered for inclusion on the Select Agent list. It is a deliberate process. The Select Agent Regulations are appropriately backwards-looking and based on a list of known agents. They are intended to protect the nation by restricting the availability of agents that are *known* from actual experience to be dangerous, that can be usefully controlled by "chain of custody" measures, and that have a high potential for biowarfare or bioterror. A list of named agents is in fact a reasonable model for the Select Agent Regulations despite the serious problems and ambiguities inherent in assigning discrete taxonomic identities to a continuum of biological organisms.

(2.) ***"Select Agent-ness" has biological and non-biological components.*** The Select Agent designation depends on a variety of considerations. Some of these are biological (such as virulence, transmissibility, dissemination, and ability to be weaponized); but others are not (such as public perception, economic impact, intelligence data, availability of countermeasures, and natural prevalence). Because the security threat posed by an agent is not determined by biological criteria alone, Select Agent status can never be predicted from sequence alone. "Select Agent" is not a scientific description; it is a policy designation.

(3.) ***Biology is not binary.*** Microorganisms are not either "potential weapons of mass destruction" or "of no concern." No single characteristic makes a microorganism a pathogen, and no clear-cut boundaries that separate a pathogen from a non-pathogen. Pathogenic microorganisms are not defined by taxonomy; it is common for a microbial species to have pathogenic and non-pathogenic members. An agent has multiple biological attributes, and the degree to which they are expressed falls along a spectrum for each biological characteristic;[6] consequently, agents pose varying degrees of risk.[7]

[6]For example, one microorganism may be highly virulent, but poorly transmissible from person to person, whereas another may spread easily but produce only mild illness.

[7]As described in Chapter 1, there are many ways to categorize microorganisms according to risk: *Biosafety in Microbiological and Biomedical Laboratories* biosafety levels 1-4; National Institutes of Health guidelines risk groups 1-4; Centers for Disease Control bioterrorism agents categories A, B, and C; and the Department of Homeland Security Bioterrorism Risk Threat Assessment and the Select Agents list, which currently is not stratified according to risk.

Moreover, the genes and sequences that could potentially be used to create a bioweapon come from all of biology. For instance, a human sequence, such as interleukin-4, could be appropriated to trigger a severe immunological response and cause illness or death. Likewise, a toxin gene from a plant could, in theory, be incorporated into a bioweapon. Microorganisms are by no means the only source of sequences of concern. Biology is diverse and dynamic and has many unclear boundaries. No single criterion or absolute threshold can be applied to identify biological threats. The biosafety framework uses several levels of containment to addresse the various degrees of risk posed by a microorganism or experiment using several levels of containment. Because of the complexity of biology, a microorganism or an experiment is best evaluated and best overseen case by case.

(4.) ***It is not feasible to predict pathogenicity from sequence now, and it will not be in the foreseeable future.*** As discussed in Chapter 2, sequence prediction in biology is subject to a hierarchy of difficulty that reflects the complexity of the system under analysis. The simplest of such predictions would probably be that of a single protein. Next in order of predictive difficulty would be a genetic pathway (a group of co-regulated multiple proteins that act in concert). The third simplest set of sequences to evaluate as a means of forecasting function are those of whole organisms alone in a controlled environment (multiple pathways act in concert). The most difficult predictive situation would be one in which two or more organisms interact in their natural environment.[8] It is that last level of complexity may give rise to the key biological attributes of pathogenicity and transmissibility, which contribute to the criteria that form the basis of inclusion of an organism on the Select Agents list.

Predicting pathogenicity, transmissibility, or environmental stability of a microorganism requires a detailed understanding of multiple attributes of the pathogen, its host, and its environment. It is a prediction problem of the greatest complexity. By the time we have a general ability to predict host-pathogen interactions on the basis of pathogen genome sequence alone, we will probably have solved a number of other major problems in biology (such as developing a vaccine for HIV, curing the common cold, and achieving personalized medicine). It might never be possible to predict pathogenicity from sequence at a level of certainty that would be required for legal statutes, such as the Select Agent Regulations, that require definitive accuracy as opposed to probabilistic risk assessment. Reliable prediction of the hazardous properties of pathogens from their genome sequence alone will require an extraordinarily detailed understanding of host, pathogen, and environment interactions integrated at the systems, organism, population, and ecosystem levels. For the foreseeable future,

[8]Consider, the enormous number of gene sequences at play and which must be choreographed as a microorganism leaves the salivary gland of a biting insect and is injected into the human tissues.

the only reliable predictor of the hazard posed by a biological agent is actual experience with it. High-level phenotypes like pathogenicity and transmissibility cannot now plausibly be predicted with the degree of certainty required for regulatory purposes, and it will probably not be possible in the foreseeable future.[9]

(5.) ***Prediction and design are linked.*** Design and prediction go hand in hand; our lack of predictive ability in biology also means that we cannot design genomes *de novo*. If we lack the ability to predict an organism's phenotype from its genome sequence, we necessarily lack the ability to design a novel genome sequence with a desired phenotype. Designing a self-replicating organism that has only to interact with simple molecules in a test tube is difficult; designing a pathogen that has to interact with a complicated host, evade the host's immune system, and be transmissible in the natural environment adds daunting layers of biological complexity. There are very few cases in which a single protein sequence has been designed to fold in a particular novel way. The first few modest successes in *de novo* design of single proteins constitute the current state of the art. Synthetic biology cannot be used to design and create an entirely novel pathogen, for exactly the same reasons that we cannot predict whether a genome sequence will be that of a pathogen. Without predictive ability, designers cannot know whether their designed sequences will work. The "entirely novel synthetic bioweapon" scenario is not plausible. However, as discussed in Chapter 3, it is possible, and even routine, to modify known organisms and to construct chimeras.

(6.) ***Synthetic genomics poses three threat scenarios that would allow a "bad actor" to obtain a pathogenic organism with Select Agent properties; one of them (modified Select Agents) is of most immediate concern.*** Chapter 3 described three scenarios in order of increasing technical difficulty, and therefore decreasing likelihood: *modified* pathogens; *chimeric* pathogens; and *designed* pathogens. More likely scenarios should be addressed before there is inordinate worry about less likely ones. The Select Agent Regulations are intended to control facile access to the most dangerous known pathogens. Synthetic genomics is beginning to make it possible to obtain pathogens by synthesis without the need for access to a live culture of an agent. A high degree of technical sophistication and great expense are necessary to synthesize and "boot" a known Select Agent genome, and an even higher degree of sophistication is required to produce a non-trivially modified

[9]It is important to note that identifying hazardous pathogens or experiments is not the same as distinguishing experiments that are legitimate from those that are illegitimate. Legitimate research aimed at understanding pathogenicity and treating infectious disease often requires work with biological hazards.

Select Agent genome (a synthetic genome derived from a Select Agent with a small number of additions, deletions, and modifications of genes) that would be likely to function; nonetheless, these are the most plausible (if unlikely) "garage laboratory" scenarios. Non-trivial chimeric constructions (more wholesale rearrangement and "assembly" of parts from different organisms into a novel whole) are extraordinarily challenging and would almost certainly require large laboratory resources and iterative optimization in an experimental testing program in susceptible hosts, contravening the Biological Weapons Convention (The committee sees the realm of chimeric genomes as beyond the regulatory scope of the Select Agent Regulations). *De novo* design remains essentially infeasible. Thus the committee believes that the most pressing issues raised in connection with the Select Agent Regulations by synthetic genomics and synthetic biology involve the synthesis or modification of known Select Agent genomes or modifications of known Select Agent genomes.

(7.) ***There is an important distinction between sequence-based* prediction *and sequence-based* classification.**

Prediction of complex biological properties is not currently feasible, just as design of an entirely novel pathogen *de novo* is not possible. For the foreseeable future, synthetic genomics and synthetic biology will be done by modification and rearrangement of genes that already exist in nature. If we assume the most plausible threat to come from modifications and rearrangements of genes from known Select Agent genomes, we can anticipate the most likely "space" of possible modifications and most obviously worrisome chimeras that might create a genome that encodes Select Agent properties. Because we can use sequence analysis to recognize genes and genomes and classify them into known families, we can use sequence analysis to designate particular genome sequences unambiguously as equivalent to "complete, infectious" Select Agents and to identify "sequences of concern."

Sequence-based classification is strictly operational—a set of tools for drawing decision boundaries around known sequences that do and do not belong to a desired classification. The tools are used now for robust and automatic classification of gene sequences into usefully annotated sequence families. For an operational definition of a complete Select Agent genome, we can define a parts list of genes that are thought to be necessary, although not sufficient, to make up a biologically functional Select Agent genome. We might even choose to simplify a classification system deliberately by defining an operationally "complete" genome as having a necessary subset of parts rather than a complete set. We should be able to establish a reasonable operational definition of the sequence space circumscribing *complete* agent genomes, as distinct from incomplete genomes or complete genomes of related non-Select Agents thus establishing a "brighter line," an unambiguous procedure for deciding when a genome sequence is assigned one of the taxonomic names on the Select Agent list.

Determining whether a sequence really does encodes a viable, functional, "infectious form" of a Select Agent is an empirical experimental question, and will long remain beyond any foreseeable predictive ability in biology. However, for the purposes of sequence-based classification, we do not need to have complete knowledge. Partial knowledge reflects the state of current knowledge, suffices for an operational definition that partitions sequence space in a way that avoids the misclassification of non-Select Agent genomes (such as those of vaccine strains or related non-pathogenic species) while trying to "deny" the spaces encompassing the modifications of Select Agent genomes that could most plausibly still encode a Select Agent pathogen.

(8.) ***Sequence-based classification could be used to address an immediate challenge raised by synthetic genomics.***

Synthetic genomics is increasingly making it possible to obtain Select Agents by synthesis rather than by access to a live laboratory culture and to create modifications that blur taxonomic classification boundaries yet still might be expected to function as a Select Agent. Because the Select Agent Regulations cover creation, transfer, and possession of complete synthetic genomes, not just those of viable Select Agents, gene and genome synthesis companies, for example, need to know unambiguously whether a customer's order is for a synthetic Select Agent genome or not. A sequence-based classification system could provide a high degree of clarity—for investigators, biohobbyists, synthesis companies, and law-enforcement officials—about what DNA sequences are subject to the Select Agent Regulations and which ones are not. The current boundaries are unclear, and this does not seem appropriate for high-consequence regulations like the Select Agent Regulations.

(9.) ***Sequence-based classification could also be used to define sequences of concern that are not themselves Select Agents, but that may nonetheless constitute a threat.***

One might argue that a disadvantage of bright line classification of Select Agent genomes is that the "bad actor" knows where the line is, too, and so can try to skirt it. What happens if the bad actor orders a Select Agent genome in pieces from different companies, or introduces just enough changes into a synthetic genome to evade Select Agent classification, or creates an entirely unexpected chimera from non-Select Agent parts?

One answer to that concern is that the Select Agent Regulations can make acquisition of Select Agents only more difficult, not impossible. It is already the case that Select Agents can be collected from the wild rather than obtained from a registered laboratory. A classification system could be designed to recognize the most plausible modified genomes and even the most obvious chimeric genomes that, according to the current state of the art, would (a) have some possibility of encoding an agent with Select Agent properties and (b) have little possibility of encoding an agent that should not be considered an Select Agent. The Select Agent list and the associated classification system would be updated as the state of the art advanced. Of course, a person with nefarious intent might be able to do better than the current state of the art in the scientific community, but this ought to be unlikely.

A second answer to the concern is that it is not and should not be the purpose of the Select Agent Regulations to regulate novel agents, any more than it is the purpose of the Select Agent Regulations to regulate access to novel emerging diseases in nature. To prohibit possession and transfer of *de novo* agents at the point of their synthesis would require the kind of forward-looking predictive system that we find infeasible. Rather, the Select Agent Regulations implement a necessarily backward-looking system based on a taxonomy of known Select Agents—already known from experience to be extraordinarily dangerous. If a new agent is found to be extraordinarily dangerous, it can be added to the Select Agent list, whether it is a naturally emerging pathogen or a synthetic. Initially, that may sound like closing the barn door after the horse is gone if we imagine a sophisticated bioterrorist engaged in designing novel agents; but it seems far more plausible that a novel agent would be discovered first as a newly emerging disease in nature or by accident in the course of legitimate biotechnology research.

The committee has a third answer. It is useful to identify suspicious sequences of concern that might be parts of a Select Agent or a bioweapon threat, even if they do not make up a complete genome subject to the Select Agents Regulations. As long as the response to a sequence of concern is flexible and does not immediately trigger regulatory or law-enforcement intervention, this can be a gray area, not a bright line. A sequence-based classification system would inherently organize and condense the current state of knowledge about the genomic composition of dangerous pathogens. The same system could be used to identify partial genomes and suspicious parts in the gray area, triggering common-sense follow-up. For example, a DNA-synthesis company might contact a customer to be sure that the customer is legitimate, and the customer knows that what is being ordered might be considered dangerous. A sequence-based classification system could help to make the identification of sequences of concern more systematic and consistent in the synthetic genomics community. Our committee referred to this as a yellow flag system (Figure 4.1, right side).

(10.) ***As predictive ability develops in biology, it will be more appropriate to use it in the context of probabilistic risk assessment (such as the yellow flag system), not in rigid classification of Select Agent properties.***

The ability to predict biological properties from genome sequence will come gradually in a long series of steps of refinement and slowly increasing accuracy. For all the reasons described in Chapter 2, it is not reasonable to

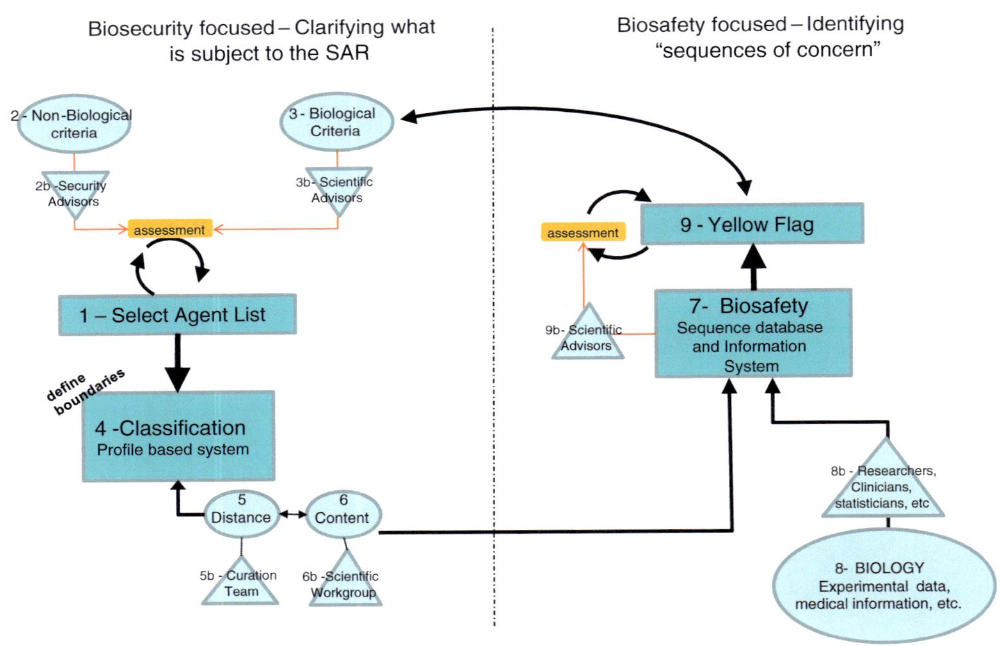

FIGURE 4.1 Concept of sequence-based classification and yellow flag systems, including differences and interactions between biosecurity and biosafety components (see also Appendix L). Black lines indicate information flow; yellow lines represent decision making. Biological (3) and non-Biological (2) criteria drive the Select Agent list (1) assessment. Profile-based classification system (4) would create sequence-based boundaries around Select Agents. Content (6) of profile-bounded sequences are input to the Biosafety system (7), which also uses experimental and medical information (8) to define Yellow Flag (9) set of sequences that indicate potential biosafety concern. Yellow flag information would inform the process of Select Agent designation. (Scientific Advisors (3b and 9b) could be carried out by a single scientific advisory panel.)

expect predictive technology to reach the accuracy necessary for defining Select Agents. However, we found it natural to think of the slowly increasing accuracy of predictive methods in the context of probabilistic risk assessment, in which the uncertainty of a prediction can be weighted appropriately. Advances in predictive technology might gradually become a counterpart of a "yellow flag" warning and biosafety framework that was initially based only on sequence-based classification. (As noted throughout this report, the classification and "yellow flag" system are presented as proposals for consideration; they should not be read as recommendations.)

The Yellow Flag System

The yellow flag system would have two primary goals: (1) to make it harder for bad actors to obtain pathogens as weapons or as tools for bioterror without detection and (2) to avoid the accidental, inadvertent, or ill-advised production of hazardous constructs by well-meaning investigators.

The yellow flag system would comprise four main elements: a centralized biosafety sequence database, annotation of the sequences as empirical evidence of the function of the genes encoded by the sequences is acquired, a process for review and assessment of the evidence to determine the disposition of the sequence of concern, and a yellow flag of the sequences that are deemed "of concern" (see Fig 4.1, right side).

There are many avenues by which a sequence might be deposited in the database and given a yellow flag, including but not restricted to the following: a researcher may observe that the gene product increases pathogenicity, the sequence may be derived from a known Select Agent and is in a region known to be critical for causing disease, or the disease-causing characteristic is eliminated when the sequence is deleted from a known pathogen. Movement of a sequence into the database can be dynamic because the system is not regulatory and a yellow flag does not restrict access to the sequence. This database system is intended to serve as a resource for information sharing.

Once a sequence is deposited in the biosafety database, it serves as a reference for anyone carrying out relevant investigations and for gene synthesis companies that would be able to compare their orders with entries in the database, screening for yellow flags. If a match occurs, the company would have a basis for notifying the purchaser of the possible concern and would request that any research results that support or refute the cause for concern be contributed to the annotations associated with the sequence in question. Similarly, other researchers carrying out experiments involving analysis of the function of yellow flag sequences would also be encouraged to provide follow-up information or references.

Scientific workgroups would be charged to analyze the annotations and make determinations as to whether the degree of concern is sufficient to merit consideration as a Select Agent, needs further study, or should be cleared of the yellow flag. A sequence may be removed from the database system entirely, although it is reasonable to retain the information in the database and indicate that the sequence has been examined and cleared. The database system would probably grow to include a variety of biosafety information, and only a subset of the sequences in the database would have yellow flags. It is important that, like the Select Agent list, the yellow flag system be fluid; sequences should be examined and yellow flags removed when they are unwarranted. The authority and resources necessary to make the process work should be provided centrally as a function supporting both biosafety and biosecurity.

We envision actions taken in response to a yellow flag as informal, prudent best practices, in that they fall outside the strict regulatory boundaries of the Select Agent Regulations. However, it would also be possible to use a yellow flag system in more formal ways. For example, an IBC or funding sponsor could ask that yellow-flagged synthetic constructs trigger special notification for purposes of oversight to track what laboratories were in possession of yellow-flagged constructs. Similarly, DNA synthesis companies might be asked to maintain records of yellow-flagged constructs that they provide to customers to facilitate forensic investigation in the event of criminal construction of a complete Select Agent from synthetic parts. Finally, a centralized system for reporting orders of yellow flag sequences could be developed to allow detection of the simplest scheme for avoiding the Select Agent Regulations—splitting the order for a viral genome or a toxin between two different gene synthesis companies.

A yellow flag biosafety system as described here would complement the Select Agent Regulations by providing oversight that is broad and flexible. It would identify sequences that potentially pose a risk without diverting attention from recognized threats or imposing restrictions and adding burden to the scientific community.

NEAR-TERM[10] MILESTONES FOR SEQUENCE-BASED CLASSIFICATION

The committee's analysis of sequence-based classification in Chapter 3 stems from a broad interpretation of its charge. However, it is the only positive and constructive response that the committee identified to address the challenges that synthetic genomics and synthetic biology pose to the Select Agent Regulations. The primary direction we were asked to consider, prediction of biological properties from sequences, is not feasible now and probably will

[10]Near-term is used here to indicate that the milestones are not dependent upon future technological advances. The technical capabilities and biological knowledge needed to achieve them are available now. Several of these milestones would improve and evolve but they could be started now, and substantial progress could be made within 5 years.

not be in the foreseeable future. The sequence-based classification discussed in Chapter 3 is technologically feasible and may improve the current system. However, such a system has limitations and potential adverse consequences.[11] Therefore, we do not specifically recommend that it be implemented. Rather, we offer the two following recommendations:

- **The sequence space around each discrete taxonomic name on the Select Agent list should be clearly defined, so that Select Agent status can be unambiguously determined from a genome sequence (for example, by a DNA synthesis company).**

  The sequence space should be broad enough to include the plausible modifications and chimeras that experts reasonably believe will probably also act as Select Agents, without encompassing existing non-Select Agents.
- **A sequence-based classification system could address this problem, and should be considered and weighed against the cost and complexity of implementing this technological augmentation to the current Select Agent Regulations.**

Specific milestones or research areas that would aid in implementing a sequence-based classification system are presented below. (Appendix L presents additional near-term milestones for consideration.)

(a.) ***A sequence database with a Select Agent focus:*** The computational sequence analysis technologies used for sequence-based classification define sequence spaces that circumscribe the known variation of sequences that are considered to belong to a useful name while excluding the known variation of sequences that are considered to be attached to different names (see Figure 3.1). A necessary precondition is to have a number of representative sequences that belong to the desired classification and a number of the most closely related sequences that do not belong. It is not sufficient to know a single representative genome sequence of each Select Agent. The more sequences that are known, the better the expected genetic variation will be understood. To provide a sound foundation for sequence-based classification of existing Select Agents, a comprehensive sequence database should be created that thoroughly covers naturally occurring genetic variation based on geographic distribution, ecological or laboratory adaptations, and those associated with clinical severity or attenuation. The database should include not only Select Agent sequences, but also a representative set of near-neighbors for each Select Agent.

[11]Including dual-use concerns, as discussed below. (See also Appendix L.)

(b.) ***An expanding sequence database of all biology:*** There are massive gaps in our knowledge of the genetic characteristics of much of the biological world. Genome and metagenome sequencing is rapidly closing some of the gaps in some groups of organisms but not others. For example, it would be useful to know much more about viral and microbial biodiversity in nature. Many new emerging pathogens (such as Nipha, SARS-CoV, and H5N1) were animal pathogens that suddenly jumped the species barrier; more sequence coverage of the viral and bacterial phylogenetic landscapes encoded in animal reservoirs would help in anticipating, monitoring, and responding quickly to future threats. Such a sequence database could be used to help to identify sequences of concern that may be appropriate to monitor in the yellow flag system in the interests of biosafety and biosecurity.

(c.) ***Define the Criteria for Select Agent Designation:*** The criteria for designating a pathogen as a Select Agent should be reviewed and clearly defined to allow unambiguous implementation of the Select Agent Regulations. The Select Agent Regulations are based in law and backed up by serious penalties. However, the criteria for designation of a pathogen as a Select Agent are not well established and include characteristics that are independent of biological or genomic characteristics. It is not always evident to the regulated community why particular agents are included on the list. Each agent that is designated as a Select Agent should have a readily justifiable reason for such designation. The criteria for Select Agent designation should be made clear and should focus on biosecurity concerns. Agents that do not meet the criteria (whether biological and non-biological) should not be added to the list. The committee recognizes that the reason for placement on the Select Agent list may involve classified information. However, even such non-biological considerations should be based on clear criteria and informed by scientific data. For instance, in some cases, it appears that past experimentation with an agent for purposes of warfare or terrorism has resulted in *de facto* inclusion on the Select Agents list. If experiments led to the conclusion that the agent is unstable, difficult to make, or poorly transmissible, then the agent might not pose a threat worthy of Select Agent designation. Furthermore, because the level of threat posed by a microorganism or toxin may change over time, (for example, countermeasures may become available or the agent may be endemic), each Select Agent and Toxin should be reevaluated regularly to ensure that it meets the criteria for Select Agent designation, and is not diverting attention from more important threats.

The committee concurs with other groups that the current system would be improved if each agent were assessed on the basis of clear criteria. Moreover, it will be difficult to create *any* clear and effective sequence based system, whether classification based or prediction based,, if the criteria and purpose of the Select Agent list remain unclear.

(d.) ***Stratification of the Select Agent list:*** The existing list of Select Agents and toxins should be reviewed on the basis of clear criteria with the goal of prioritizing the Select Agent list on the basis of risk. Mechanisms for timely inclusion *and* removal of an agent or toxin from the Select Agent list are necessary for a robust oversight system. Several recent advisory panels have recommended stratification or reduction of the Select Agent list, and we are in agreement with their recommendations.[12,13] As stated in the 2009 National Research Council report, "a list of more than 80 agents of varying risks dilutes attention from those that pose the greatest degree of concern, which may, in the process, render the nation less secure. It would be more effective to focus the highest scrutiny on those agents that are, indeed, of greatest concern . . . (NRC 2009a)" A gene sequence based classification system is certainly an example of this situation.[14] Classifying the current 82 Select Agents would require 82 parts lists and several thousand profiles for the parts, and, as mentioned, each Select Agent classification would need to be carefully tested and maintained. A classification system would require a small team of full-time staff to develop and maintain. Sequence curation would require substantial work. Prioritizing the Select Agent list *on the basis of risk* would make *any* sequence-based approach to oversight more feasible

LONG-TERM MILESTONES FOR GENOME SEQUENCE-BASED SELECT AGENT REGULATIONS

The use of the term *milestones* may be somewhat misleading here, in as much as the research described is ongoing, and will evolve in a continuous and

[12]"The list should be either reduced or stratified so that biosecurity measures can be more easily applied by the registered entities according to the level of risk" and "Perform a risk assessment for each select agent and toxin on the BSAT list and develop a stratification scheme that includes biodefense and biosecurity criteria, as well as risk to public health, so that security measures may be implemented based upon risk" (NSABB 2009b). Report of the Working Group on Strengthening the Biosecurity of the United States.

[13]"RECOMMENDATION 3: The list of select agents and toxins should be stratified in risk groups according to the potential use of the agent as a biothreat agent, with regulatory requirements and procedures calibrated against such stratification. Importantly, mechanisms for timely inclusion or removal of an agent or toxin from the list are necessary and should be developed (NRC 2009b)."

[14]The committee wants to be clear that implementation of a classification system is not a reason to subtract specific agents from or add specific agents to the Select Agent list. Rather, implementation of a sequence based system is a benefit of reducing the list.

interrelated way. A robust oversight system will have to be able to evolve as well, with continuing integration of scientific advancements. The milestones toward developing the knowledge and capabilities needed to enable a predictive oversight system (or to enhance a classification system) are shared among all fields of biology. It is a major goal of all biology to understand how DNA sequence determines the properties of biological systems, ranging upwards in complexity from single macromolecules to pathways, organisms, populations, and ecosystems. We are far from that goal. Successes in prediction and design at each level of complexity in biology as a whole are the relevant achievements to watch for, before we will be able to predict confidently from genome sequence analysis how a designed organism would replicate, interact with a host, evade a host immune system, and spread in a population to cause disease.

The goal of a predictive oversight system is so far out in front of current biological understanding that it would be unwise to attempt to address it in detail. Instead, we offer the following general milestones:

- **Ability to predict accurately the function of individual proteins from genome sequence sequence, including what ligands or macromolecules they bind to, what reactions they catalyze, where they localize, and what the kinetic rate constants for these processes are.**
- **Ability to predict accurately from genome sequence the output of biochemical, regulatory, and genetic pathways (modules) of several proteins acting together.**
- **Ability to predict accurately the behavior of a whole organism from its genome sequence.**
- **Ability to predict accurately from their genome sequences the interactions of organisms in their natural environment from their genome sequences, such as microbe-host symbioses or host-pathogen interactions.**

Those very general goals are already shared by all the biomedical sciences for advancing understanding of all biological systems. They are not peculiar to Select Agents, or even to infectious disease.

Although specific milestones for prediction of Select Agents are far beyond current scientific insight, the committee is able to identify promising research areas and technologies that would improve the ability to predict gene function, enhance understanding of infectious disease, and consequently strengthen biosecurity.[15] What follows is not intended to be an exhaustive list; research

[15]This is consistent with the National Strategy for countering Biological Threats "The objectives of our Strategy [include] . . . Promote global health security: Activities that should be taken to increase the availability of and access to knowledge and products of the life sciences that can help reduce impacts of outbreaks of infectious disease whether of natural, accidental, or deliberate origin." NSC (2009). *National strategy for countering biological threats.* Washington, D.C., National Security Council.

findings in fields not described could well provide important advances in our understanding of genotype-to-phenotype prediction.

The committee recommends supporting these research efforts and technological developments, with the understanding that predicting function from sequence is a major biological goal. Progress in these efforts could be applied to strengthen a gene sequence-based oversight system as it evolves, but, the value of the research extends far beyond its potential contribution to biosecurity.

(a.) ***Protein structure and function:*** There are important gaps in our understanding of the relationships between nucleic acid sequence and protein structure, and between protein structure and gene function. Developing a better understanding of the relationships between nucleic acid sequence, protein structure, and gene function will be critical for improving our knowledge base.

(b.) ***Gene expression and regulation:*** Gene function may be multifactorial—based on interactions with other genes, physiological conditions, and other regulatory events. Developing a better understanding of factors that regulate gene functions is needed. If an organism has specialized gene products for its virulence, it must be able to use them when they are needed but not squander its metabolic energy in producing them aimlessly or risk having them detected by host defenses and prematurely neutralized. Consequently, regulating the expression of virulence factors is an additional, essential complication of a pathogenic microorganism's life. The number of well-characterized virulence regulatory systems is increasing rapidly, in part because of the development of rapid methods for screening gene expression on a genome-wide basis (for example, with DNA microarrays). At the same time, relatively little is known about both the specific environmental signals to which the systems respond and the exact role of the responses in the course of human infection.

(c.) ***Pathogenic mechanisms:*** The molecular basis of the pathogenic characteristics of currently designated Select Agents is, in general, poorly understood, may be multigenic, may in some cases be greatly influenced by one or a few single-nucleotide polymorphisms, and may be regulated by mechanisms that are not well defined. The molecular basis of novel pathogens or human-made organisms with pathogenic potential is also not established. To inform a gene sequence-based classification system, improve our biodefense capabilities, and, most important, combat infectious disease and improve public health, a better understanding of the molecular basis of virulence should be developed. Pathogenesis due to an existing Select Agent or a novel pathogen is often host-specific, but there is little information to explain the contrast between pathogenesis in a receptive host species and the absence of pathogenesis in a species (or individual) not affected by the pathogen. Any determination of the molecular basis of the pathogenic characteristics of a microorganism must include consideration of its host and the host response. Developing a comprehensive understanding of the pathogen-host interactions that result in the creation of a disease state would be an important achievement.

(d.) ***Animal models of disease:*** For many Select Agents, there are no surrogate experimental hosts for characterizing virulence; the only suitable host for a human pathogen may be humans. An adequate understanding of the function of a gene or cluster of genes cannot be obtained through computational modeling alone; to ensure confidence in results, it is essential in determining virulence characteristics should include experimental validation of function in an appropriate model system. Further development of genetically characterized animal models of various species, including non-human primates, is an important objective. For instance, current efforts to create the "Collaborative Cross" and related genetically well-defined and well-characterized mice will provide a valuable new tool to assist in the understanding of host-pathogen interactions. Novel model systems that more closely replicate human disease processes—such as humanized mice, in vitro models of human organ systems, and complete *in silico* models that recapitulate human physiological processes at a molecular level—are needed.

(e.) ***Data and information management for Systems Biology:*** A dynamic, sequence-based program will require creation of massive new and well-integrated databases to manage greatly expanded sequence information on orders and families of organisms yet to be examined; enumeration of protein-fold families; host pathways; protein structural determinations, including posttranslational modifications; the genetic basis of virulence and immune response from the perspective of the host and the pathogen at both the pathway interaction and more detailed 3-D structural interaction levels; and vastly improved software capabilities to use the databases to predict 3-D structural effects of nucleic acid variations and host interactions accurately, especially in relation to pathogenic effects.

(f.) ***Synthetic Biology:*** Synthetic biology approaches biology from an engineering perspective; it is aimed at solving a problem, creating tools, and designing or improving a system. All existing and reasonably foreseeable uses of synthetic biology involve modification or rearrangement of existing biological components. For instance, a precursor of the antimalarial compound artemisinin is being produced in *E. coli*, and other microorganisms are being designed to address biofuel production. The design of such pathways and chimeras is no easy task, and the entirely *de novo* design of genomes and organisms remains science fiction. That is due largely to the difficulties in predicting function from sequences, as described in Chapter 2; biological context is key to gene or protein function. As discussed in the recent *Nature News* feature "Five Hard Truths for Synthetic Biology," the developing field of synthetic biology faces several important challenges.[16] They are centered around translating biological complexity into simple tools and standardized parts that behave in a predictable ways. The committee is in agreement with the National Science Advisory Board for Biosecurity, which has stated that "synthetic biology is a rapidly evolving field, and, given its potential benefits to public health and national and economic security, research in these disciplines should be encouraged and maintained."

(g.) ***Metagenomics (phylogenomics):*** Environmental metagenomic sequencing of soils, seawater, and other complex samples consistently yields a high percentage of proteins of unknown function. It is clear that many natural offensive and defensive mechanisms that may have relevance to furthering our understanding of human pathogenesis await discovery. The advent of short-read sequencing technologies is making deep studies of complex environmental samples possible. The flood of data resulting from such studies is illustrating the need for better computational tools and infrastructure to manage, analyze, and correlate staggering amounts of information. Such efforts should be strongly supported inasmuch as unexpected discoveries from unknown organisms may prove to yield more advances than incremental hypotheses related to known organisms.

(h.) ***Microbiome:*** Although possibly germ-free (gnotobiotic) before birth, humans develop a resident microbiota shortly after birth. The human microbiome is the subject of intensive study, including the major international Human Microbiome Project (HMP). Because of advances in DNA sequencing technologies and improvements in bioinformatics, it has become possible to characterize the great diversity in the human microbiota. In 2007, the National Institutes of Health launched

[16] "Many of the parts are undefined"; "The circuitry is unpredictable"; "The complexity is unwieldy"; "Many parts are incompatible"; and "Variability crashes the system" (Kwok 2010).

the Human Microbiome Project (HMP) as one of its major roadmap initiatives. This major scientific endeavor has the following aims:

- Determining whether individuals share a core human microbiome.
- Understanding whether changes in the human microbiome can be correlated with changes in human health.
- Developing the technological tools to support these goals.
- Addressing the ethical, legal, and social complications raised by human microbiome research.

The Human Microbiome Project will add an enormous amount of additional microbial sequence to the already burgeoning database. That will be invaluable as we continue to sort out the sequences that have real predictive value instead of being merely suggestive because of some degree of relative homology with a putative virulence factor of a pathogen and especially of a Select Agent.

CONCLUSION

The milestones and focus areas listed above aim either to expand the general frontiers of biological knowledge, or to apply existing knowledge to the Select Agent Regulations. Our committee was deeply uncomfortable with research programs that would seek to expand knowledge solely for the purposes of improving the Select Agent Regulations.

Developing the ability to predict Select Agent pathogenicity from genome sequence raises serious dual-use concerns because prediction and design go hand in hand. Accurate computational prediction of Select Agents from genome sequences enables computational design and optimization of bioweapon genome sequences. Two major goals of biology are to predict phenotype from genotype and to improve public health by understanding pathogenicity. It does not seem wise to make *special* plans for an effort in predicting the characteristics of Select Agents, in advance of other important frontiers of biological knowledge.

It is more prudent to base the Select Agent Regulations on the current state of biological knowledge, as an applied problem, not a basic research problem. Predictive successes in the general biology research community should be passively monitored. Once biology in general approached the goal of determining pathogenicity from sequence, it would be appropriate to consider a predictive oversight system to identify Select Agent properties accurately from a novel genome sequence. That time may not come for decades, and it may be more than a century away.

And in the meantime? The technology and knowledge base for sequence-based classification exist now, as we described in Chapter 3. Even a classification system can present dual-use issues, in that for the system to be usefully implemented, the information must be shared. Listing the parts of a Select Agent and identifying other sequences of concern entirely on the basis of their potential to be dangerous when incorporated into a synthetic construct disseminates knowledge that theoretically could facilitate the design of a synthetic pathogen by a bad actor. However, inasmuch as this knowledge would be based on the current published state of the art (and pathogen sequences that are already widely available in GenBank), any additional dual-use concerns are not nearly as grave.

The Select Agent Regulations strive to balance a need for regulating access to the most dangerous pathogens with minimizing regulatory burdens on basic biological research aimed at monitoring, understanding, treating, and preventing disease. If the Select Agent Regulations are too burdensome, they may diminish long-term safety. Our report stops short of recommending the implementation of any specific sequence-based system for defining Select Agents; it was not our charge, and we were not properly constituted to estimate the costs, benefits, or risks associated with any specific implementation. We do find that the sequence-based classification system and yellow flag system of Chapter 3 are technologically feasible, but we have not carefully examined their costs or their effects on basic research or national security (see Appendix L). We have made no argument that the favorable aspects of using such systems to clarify a sequence-based definition of the discrete taxonomic names on the Select Agent list would outweigh any adverse aspects of creating additional layers of complexity in the regulatory framework. **Rather, our principal finding is that sequence-based *prediction* of Select Agent properties is not feasible and is unlikely to be feasible in the foreseeable future; any research effort dedicated *solely* to this purpose is likely to have *only* adverse consequences.**

4

References

Abramovitch, R. B., J. C. Anderson, et al. (2006). "Bacterial elicitation and evasion of plant innate immunity." *Nat Rev Mol Cell Biol* **7**(8): 601-611.

Afonnikov, D. A. and N. A. Kolchanov (2004). "CRASP: a program for analysis of coordinated substitutions in multiple alignments of protein sequences." *Nucl. Acids Res.* **32**(suppl_2): W64-68.

Agarwal, K. L., H. Buchi, et al. (1970). "Total Synthesis of the Gene for an Alanine Transfer Ribonucleic Acid from Yeast." *Nature* **227**(5253): 27-34.

Agarwal, R., T. Binz, et al. (2005). "Analysis of Active Site Residues of Botulinum Neurotoxin E by Mutational, Functional, and Structural Studies: Glu335Gln Is an Apoenzyme." *Biochemistry* **44**(23): 8291-8302.

Alcami, A. (2003). Viral mimicry of cytokines, chemokines and their receptors. *Nature Reviews Immunology*, Nature Publishing Group. **3:** 36.

Anandalakshmi, R., G. J. Pruss, et al. (1998). "A viral suppressor of gene silencing in plants." *Proceedings of the National Academy of Sciences of the United States of America* **95**(22): 13079-13084.

Ben-David, M., O. Noivirt-Brik, et al. (2009). "Assessment of CASP8 structure predictions for template free targets." *Proteins: Structure, Function, and Bioinformatics* **77**(S9): 50-65.

Bent, A. F. and D. Mackey (2007). "Elicitors, Effectors, and R Genes: The New Paradigm and a Lifetime Supply of Questions." *Annual Review of Phytopathology* **45**(1): 399-436.

Berger, K. M., W. Pinard, et al. (2009). Minimizing the Risks of Synthetic DNA: Scientists' Views on the U.S. Government's Guidance on Synthetic Genomics.

Black, D. S. and J. B. Bliska (2000). "The RhoGAP activity of the *Yersinia pseudotuberculosis* cytotoxin YopE is required for antiphagocytic function and virulence." *Molecular Microbiology* **37**(3): 515-527.

Bowie, A. G. and L. Unterholzner (2008). "Viral evasion and subversion of pattern-recognition receptor signalling." *Nat Rev Immunol* **8**(12): 911-922.

Breeze, Budowle, et al., Eds. (2005). *Microbial Forensics*, Elsevier Academic Press.

Brigneti, G., O. Voinnet, et al. (1998). "Viral pathogenicity determinants are suppressors of transgene silencing in Nicotiana benthamiana." *EMBO J* **17**(22): 6739-6746.

Buller, R. M. and G. J. Palumbo (1991). "Poxvirus pathogenesis." *Microbiol. Mol. Biol. Rev.* **55**(1): 80-122.

Casadevall, A. and D. A. Relman (2010). "Microbial threat lists: obstacles in the quest for biosecurity?" *Nature Reviews Microbiology* **8**(2): 149-154.

CDC (1974). Classification of Etiologic Agents on the Basis of Hazard. U. S. N. C. D. Center. Atlanta, GA.

CDC (2005). Notification of Exclusion of Attenuated Strains.

CDC/NIH. (2007). "Biosafety in microbiological and biomedical laboratories." from http://purl.access.gpo.gov/GPO/LPS121160.

Cello, J., A. V. Paul, et al. (2002). "Chemical Synthesis of Poliovirus cDNA: Generation of Infectious Virus in the Absence of Natural Template." *Science* **297**(5583): 1016-1018.

Chang, J. H., J. M. Urbach, et al. (2005). "A High-Throughput, Near-Saturating Screen for Type III Effector Genes from Pseudomonas syringae." *Proceedings of the National Academy of Sciences of the United States of America* **102**(7): 2549-2554.

Chang, M. C. Y., R. A. Eachus, et al. (2007). "Engineering Escherichia coli for production of functionalized terpenoids using plant P450s." *Nat Chem Biol* **3**(5): 274-277.

Commission on the Prevention of WMD Proliferation and Terrorism. (2008). *World at Risk : the report of the Commission on the Prevention of WMD Proliferation and Terrorism*. New York, Vintage Books.

Couch, B. C., I. Fudal, et al. (2005). "Origins of Host-Specific Populations of the Blast Pathogen Magnaporthe oryzae in Crop Domestication With Subsequent Expansion of Pandemic Clones on Rice and Weeds of Rice." *Genetics* **170**(2): 613-630.

Damon, I. K. (2007). Genus Orthopoxvirus: Variola virus. *Poxviruses*: 47-64.

DHHS (1979). "Interstate shipment of etiologic agents; proposed rule." *Federal Register* **44**(226): 66853-5.

DHHS (1996). 42 CFR Part 72: Additional Requirements for Facilities Transferring or Receiving Select Agents, Federal Register. **61:** 55190-55200.

DHHS (2002). Public Health Security and Bioterrorism Preparedness and Response Act of 2002.

DHHS (2005). 42 CFR 72 and 73 and 42 CFR Part 1003: Possession, Use, and Transfer of Select Agents and Toxins; Final Rule, Federal Register. **70:** 12294-13325.

DHHS (2009). Screening Framework Guidance for Synthetic Double-Stranded DNA Providers, Federal Register. **74**.

DHS (2006). Bioterrorism Risk Assessment. Fort Detrick, MD, Biological Threat Characterization Center of the National Biodefense Analysis and Countermeasures Center.

Dias, M. B., L. Reyes-Gonzalez, et al. (2010). "Effects of the USA PATRIOT ACT and the 2002 Bioterrorism Preparedness Act on select agent research in the United States." *Proceedings of the National Academy of Sciences* **107**: 9556-9561.

Doolittle, R. (1981). "Similar amino acid sequences: chance or common ancestry?" *Science* **214**(4517): 149-159.

Dunoyer, P., C.-H. Lecellier, et al. (2004). "Probing the MicroRNA and Small Interfering RNA Pathways with Virus-Encoded Suppressors of RNA Silencing." *Plant Cell* **16**(5): 1235-1250.

Dymond, J. S., L. Z. Scheifele, et al. (2009). "Teaching Synthetic Biology, Bioinformatics and Engineering to Undergraduates: The Interdisciplinary Build-a-Genome Course." *Genetics* **181**(1): 13-21.

Esposito, J. J., S. A. Sammons, et al. (2006). "Genome Sequence Diversity and Clues to the Evolution of Variola (Smallpox) Virus." *Science*: 1125134.

Fallman, M., K. Andersson, et al. (1995). "Yersinia pseudotuberculosis inhibits Fc receptor-mediated phagocytosis in J774 cells." *Infection and Immunity* **63**(8): 3117-3124.

Fishman, J. A. (2007). "Infection in Solid-Organ Transplant Recipients." *N Engl J Med* **357**(25): 2601-2614.

Flannagan, R. S., G. Cosio, et al. (2009). "Antimicrobial mechanisms of phagocytes and bacterial evasion strategies." *Nat Rev Micro* **7**(5): 355-366.

Fleishman, S. J., O. Yifrach, et al. (2004). "An evolutionarily conserved network of amino acids mediates gating in voltage-dependent potassium channels." *Journal of Molecular Biology* **340**(2): 307-318.

Frankel, A., P. Welsh, et al. (1990). "Role of arginine 180 and glutamic acid 177 of ricin toxin A chain in enzymatic inactivation of ribosomes." *Mol. Cell. Biol.* **10**(12): 6257-6263.

Fujii, N., K. Kimura, et al. (1992). "A zinc-protease specific domain in botulinum and tetanus neurotoxins." *Toxicon* **30**(11): 1486-1488.

Genomesonline. (2009). "GOLD: Genomes OnLine Databases.", from http://www.genomesonline.org/gold.cgi/.

Gibson, D. G., J. I. Glass, et al. (2010). "Creation of a Bacterial Cell Controlled by a Chemically Synthesized Genome." *Science* **329**(5987): 52-56.

Gillespie, J. J., N. C. Ammerman, et al. (2009). "Louse- and flea-borne rickettsioses: biological and genomic analyses." *Vet. Res.* **40**(2).

Graff, J. W., K. Ettayebi, et al. (2009). "Rotavirus NSP1 Inhibits NFÎºB Activation by Inducing Proteasome-Dependent Degradation of Î²-TrCP: A Novel Mechanism of IFN Antagonism." *PLoS Pathog* **5**(1): e1000280.

Gubser, C., S. Hue, et al. (2004). "Poxvirus genomes: a phylogenetic analysis." *J Gen Virol* **85**(1): 105-117.

Haber, E. and C. B. Anfinsen (1962). "Side-chain Interactions Governing the Pairing of Half-cystine Residues in Ribonuclease." *Journal of Biological Chemistry* **237**(6): 1839-1844.

Hannenhalli, S. S. and R. B. Russell (2000). "Analysis and prediction of functional sub-types from protein sequence alignments." *Journal of Molecular Biology* **303**(1): 61-76.

He, Y., D. C. Yeh, et al. (2005). "Solution NMR Structures of IgG Binding Domains with Artificially Evolved High Levels of Sequence Identity but Different Folds." *Biochemistry* **44**(43): 14055-14061.

Hussain, S., S. Perlman, et al. (2008). "Severe Acute Respiratory Syndrome Coronavirus Protein 6 Accelerates Murine Hepatitis Virus Infections by More than One Mechanism." *J. Virol.* **82**(14): 7212-7222.

Iyer, L. M., S. Balaji, et al. (2006). "Evolutionary genomics of nucleo-cytoplasmic large DNA viruses." *Virus Research* **117**(1): 156-184.

Janin, J. (2005). "Assessing predictions of protein-protein interaction: The CAPRI experiment." *Protein Science* **14**(2): 278-283.

Joiner, K., E. Brown, et al. (1983). "Studies on the mechanism of bacterial resistance to complement-mediated killing. III. C5b-9 deposits stably on rough and type 7 S. pneumoniae without causing bacterial killing." *J Immunol* **130**(2): 845-849.

Keedy, D. A., C. J. Williams, et al. (2009). "The other 90% of the protein: Assessment beyond the Calphas for CASP8 template-based and high-accuracy models." *Proteins: Structure, Function, and Bioinformatics* **77**(S9): 29-49.

Kim, D. W., G. Lenzen, et al. (2005). "The Shigella flexneri effector OspG interferes with innate immune responses by targeting ubiquitin-conjugating enzymes." *Proceedings of the National Academy of Sciences of the United States of America* **102**(39): 14046-14051.

Kim, Y., D. Misna, et al. (1992). "Structure of a ricin mutant showing rescue of activity by a non-catalytic residue." *Biochemistry* **31**(12): 3294-3296.

Kingsley, R. A. and A. J. Bäumler (2000). "Host adaptation and the emergence of infectious disease: the *Salmonella* paradigm." *Molecular Microbiology* **36**(5): 1006-1014.

Konkel, M. E. and K. Tilly (2000). "Temperature-regulated expression of bacterial virulence genes." *Microbes and Infection* **2**(2): 157-166.

Kortepeter, M. G. and G. W. Parker (1999). "Potential biological weapons threats." *Emerging infectious diseases* **5**(4).

Kuhlman, B., G. Dantas, et al. (2003). "Design of a Novel Globular Protein Fold with Atomic-Level Accuracy." *Science* **302**(5649): 1364-1368.

Kwok, R. (2010). "Five hard truths for synthetic biology." *Nature.* **463**(7279): 288.
Lacy, D. B., W. Tepp, et al. (1998). "Crystal structure of botulinum neurotoxin type A and implications for toxicity." *Nat Struct Biol* **5**(10): 898-902.
Lambris, J. D., D. Ricklin, et al. (2008). "Complement evasion by human pathogens." *Nat Rev Micro* **6**(2): 132-142.
Lartigue, C., S. Vashee, et al. (2009). "Creating Bacterial Strains from Genomes That Have Been Cloned and Engineered in Yeast." *Science* **325**(5948): 1693-1696.
Lee, S. K., H. Chou, et al. (2008). "Metabolic engineering of microorganisms for biofuels production: from bugs to synthetic biology to fuels." *Current Opinion in Biotechnology* **19**(6): 556-563.
Lefkowitz, E. J., C. Wang, et al. (2006). "Poxviruses: past, present and future." *Virus Research* **117**(1): 105-118.
Li, Y., D. S. Carroll, et al. (2007). "On the origin of smallpox: Correlating variola phylogenics with historical smallpox records." *Proceedings of the National Academy of Sciences* **104**(40): 15787-15792.
Lindesmith, L., C. Moe, et al. (2003). "Human susceptibility and resistance to Norwalk virus infection." *Nat Med* **9**(5): 548-553.
López, G., I. Ezkurdia, et al. (2009). "Assessment of ligand binding residue predictions in CASP8." *Proteins: Structure, Function, and Bioinformatics* **77**(S9): 138-146.
Lu, R., A. Folimonov, et al. (2004). "Three distinct suppressors of RNA silencing encoded by a 20-kb viral RNA genome." *Proceedings of the National Academy of Sciences of the United States of America* **101**(44): 15742-15747.
Matson, J. S., J. H. Withey, et al. (2007). "Regulatory Networks Controlling Vibrio cholerae Virulence Gene Expression." *Infect. Immun.* **75**(12): 5542-5549.
Maurelli, A. T., R. E. Fernandez, et al. (1998). ""Black holes" and bacterial pathogenicity: A large genomic deletion that enhances the virulence of Shigella spp. and enteroinvasive Escherichia coli." *Proceedings of the National Academy of Sciences of the United States of America* **95**(7): 3943-3948.
Maurer, S. M., M. Fischer, et al. (2009). Making Commercial Biology Safer: What the Gene Synthesis Industry Has Learned About Screening Customers and Orders.
McAuley, J. L., K. Zhang, et al. "The Effects of Influenza A Virus PB1-F2 Protein on Polymerase Activity Are Strain Specific and Do Not Impact Pathogenesis." *J. Virol.* **84**(1): 558-564.
McLeod, S. M., H. H. Kimsey, et al. (2005). "CTXφ; and *Vibrio cholerae*: exploring a newly recognized type of phage-host cell relationship." *Molecular Microbiology* **57**(2): 347-356.
Mohamed, M. R., M. M. Rahman, et al. (2009). "Proteomic screening of variola virus reveals a unique NF-Î°B inhibitor that is highly conserved among pathogenic orthopoxviruses." *Proceedings of the National Academy of Sciences* **106**(22): 9045-9050.
Moss, B. (2007). Poxviridae: The viruses and their replication. *Fields Viology*. D. M. Knipe and P. M. Howley, Lippincott, William and Wilkins: 2905-2946.
Moult, J. (2005). "A decade of CASP: progress, bottlenecks and prognosis in protein structure prediction." *Current Opinion in Structural Biology* **15**(3): 285-289.
Moulton, E. A., J. P. Atkinson, et al. (2008). "Surviving Mousepox Infection Requires the Complement System." *PLoS Pathog* **4**(12): e1000249.
Munishkin, A. and I. G. Wool (1995). "Systematic Deletion Analysis of Ricin A-Chain Function." *Journal of Biological Chemistry* **270**(51): 30581-30587.
NBACC (2009). National Biodefense Analysis & Countermeasures Center Strategic Plan. NBACC.
Neish, A. (2004). "Bacterial inhibition of eukaryotic pro-inflammatory pathways." *Immunologic Research* **29**(1): 175-185.
NIH (1976). *Guidelines for research involving recombinant DNA molecules : NIH guidelines*, U.S. Dept. of Health, Education, and Welfare, Public Health Service.

NRC (2004). *Biotechnology research in an age of terrorism*. Washington, D.C., National Academies Press.

NRC (2008). *Department of Homeland Security bioterrorism risk assessment: a call for change*. Washington, D. C., National Academies Press.

NRC (2009a). *Live Variola virus: Considerations for Continuing Research*. Washington, DC, National Academies Press.

NRC (2009b). *Responsible Research with Biological Select Agents and Toxins*. Washington, DC, National Academies Press.

NRC (2009c). *A survey of attitudes and actions on dual use research in the life sciences : a collaborative effort of the National Research Council and the American Association for the Advancement of Science*. Washington, D.C., National Academies Press.

NSABB (2006) "Addressing Biosecurity Concerns Related to the Synthesis of Select Agents."

NSABB (2009a). Enhancing Personnel Reliability among Individuals with Access to Select Agents.

NSABB (2009b). Report of the Working Group on Strengthening the Biosecurity of the United States.

NSC (2009). *National strategy for countering biological threats*. Washington, D.C., National Security Council.

Odom, M. R., R. Curtis Hendrickson, et al. (2009). "Poxvirus protein evolution: Family wide assessment of possible horizontal gene transfer events." *Virus Research* **144**(1-2): 233-249.

OECD (Organisation for Economic Co-operation and Development) (2007). *OECD Best Practice Guidelines on Biosecurity for BRCs (Biological Resource Centers)*. Paris: OECD. Available at ,http://www.oecd.org/dataoecd/6/27/38778261.pdf>.

Park, I. H., D. G. Pritchard, et al. (2007). "Discovery of a New Capsular Serotype (6C) within Serogroup 6 of Streptococcus pneumoniae." *J. Clin. Microbiol.* **45**(4): 1225-1233.

Paya, C. V. (1993). "Fungal Infections in Solid-Organ Transplantation." *Clinical Infectious Diseases* **16**(5): 677-688.

Pesenti, P. T. (2009). DHS S&T Bioterrorism Risk Assessment (BTRA).

Preston, R. (1998). Annals of Warfare: The Bioweaponeers. *The New Yorker*: 52-65.

Qian, B., S. Raman, et al. (2007). "High-resolution structure prediction and the crystallographic phase problem." *Nature* **450**(7167): 259-264.

Qiu, W., J.-W. Park, et al. (2007). "Tombusvirus P19-Mediated Suppression of Virus-Induced Gene Silencing Is Controlled by Genetic and Dosage Features That Influence Pathogenicity." *Molecular Plant-Microbe Interactions* **15**(3): 269-280.

Raman, S., R. Vernon, et al. (2009). "Structure prediction for CASP8 with all-atom refinement using Rosetta." *Proteins: Structure, Function, and Bioinformatics* **77**(S9): 89-99.

Read, T. D., S. N. Peterson, et al. (2003). "The genome sequence of Bacillus anthracis Ames and comparison to closely related bacteria." *Nature* **423**(6935): 81-86.

Rotz, L. D., A. S. Khan, et al. (2002). "Public health assessment of potential biological terrorism agents." *Emerging infectious diseases* **8**(2): 225-30.

Schnoes, A. M., S. D. Brown, et al. (2009). "Annotation Error in Public Databases: Misannotation of Molecular Function in Enzyme Superfamilies." *PLoS Comput Biol* **5**(12): e1000605.

Schorey, J. S., M. C. Carroll, et al. (1997). "A Macrophage Invasion Mechanism of Pathogenic Mycobacteria." *Science* **277**(5329): 1091-1093.

Schwan, T. G. and J. Piesman (2002). "Vector Interactions and Molecular Adaptations of Lyme Disease and Relapsing Fever Spirochetes Associated with Transmission by Ticks." *Emerg Infect Dis* **8**(2): 115-21.

Sing, A., D. Reithmeier-Rost, et al. (2005). "A hypervariable N-terminal region of Yersinia LcrV determines Toll-like receptor 2-mediated IL-10 induction and mouse virulence." *Proceedings of the National Academy of Sciences of the United States of America* **102**(44): 16049-16054.

Sodhi, A., S. Montaner, et al. (2004). "Viral hijacking of G-protein-coupled-receptor signalling networks." *Nat Rev Mol Cell Biol* **5**(12): 998-1012.

Soosaar, J. L. M., T. M. Burch-Smith, et al. (2005). "Mechanisms of plant resistance to viruses." *Nat Rev Micro* **3**(10): 789-798.

Stukenbrock, E. H., S. Banke, et al. (2007). "Origin and Domestication of the Fungal Wheat Pathogen Mycosphaerella graminicola via Sympatric Speciation." *Mol Biol Evol* **24**(2): 398-411.

Stukenbrock, E. H. and B. A. McDonald (2008). "The Origins of Plant Pathogens in Agro-Ecosystems." *Annual Review of Phytopathology* **46**(1): 75-100.

Sumby, P., S. Zhang, et al. (2008). "A Chemokine-Degrading Extracellular Protease Made by Group A Streptococcus Alters Pathogenesis by Enhancing Evasion of the Innate Immune Response." *Infect. Immun.*: IAI.01354-07.

Sutton, V. (2009). "Survey Finds Biodefense Researcher Anxiety - Over Inadvertently Violating Regulations." *Biosecurity and Bioterrorism: Biodefense Strategy, Practice, and Science* **7**(2).

The White House. (2004). HSPD-10: Biodefense in the 21st Century.

Tscharke, D. C., P. C. Reading, et al. (2002). "Dermal infection with vaccinia virus reveals roles for virus proteins not seen using other inoculation routes." *J Gen Virol* **83**(8): 1977-1986.

Tucker, J. (2003). "Preventing the Misuse of Pathogens: The Need for Global Biosecurity." *Arms Control Today* **33**.

Upton, C., S. Slack, et al. (2003). "Poxvirus Orthologous Clusters: toward Defining the Minimum Essential Poxvirus Genome." *J. Virol.* **77**(13): 7590-7600.

van der Does, H. C. and M. Rep (2007). "Virulence Genes and the Evolution of Host Specificity in Plant-Pathogenic Fungi." *Molecular Plant-Microbe Interactions* **20**(10): 1175-1182.

Vivian, A. and M. J. Gibbon (1997). "Avirulence genes in plant-pathogenic bacteria: signals or weapons?" *Microbiology* **143**(3): 693-704.

Voinnet, O., C. Lederer, et al. (2000). "A Viral Movement Protein Prevents Spread of the Gene Silencing Signal in Nicotiana benthamiana." *Cell* **103**(1): 157-167.

Walport, M. J. (2001a). "Complement- First of Two Parts." *N Engl J Med* **344**(14): 1058-1066.

Walport, M. J. (2001b). "Complement- Second of Two Parts." *N Engl J Med* **344**(15): 1140-1144.

WHO (World Health Organization) (2004). *Laboratory Biosafety Manual*, 3rd ed. WHO/CDS/CSR/LYO/2004.11 Geneva: WHO. Available at <http://www.who.int/csr/resources.publications/biosafety/WHO_CDS_CSR_LYO_2004_11/en/>.

Working Group on Strengthening the Biosecurity of the United States. (2009). Report of the Working Group on Strengthening the Biosecurity of the United States.

Wu, K., W. Li, et al. (2009). "Crystal structure of NL63 respiratory coronavirus receptor-binding domain complexed with its human receptor." *Proceedings of the National Academy of Sciences* **106**(47): 19970-19974.

Yoshida, N., K. Oeda, et al. (2001). "Protein function: Chaperonin turned insect toxin." *Nature* **411**(6833): 44-44.

Zhang, J., Q. Wang, et al. "MUFOLD: A new solution for protein 3D structure prediction." *Proteins: Structure, Function, and Bioinformatics* **78**(5): 1137-1152.

Zychlinsky, A., B. Kenny, et al. (1994). "IpaB mediates macrophage apoptosis induced by *Shigella flexneri*." *Molecular Microbiology* **11**(4): 619-627.

Appendix A

Statement of Task

NIH has requested the National Research Council to convene an ad hoc committee to identify the scientific advances that would be necessary to permit serious consideration of developing and implementing an oversight system for select agents that is based on predicted features and properties encoded by nucleic acids rather than a relatively static list of specific agents and taxonomic definitions.

The committee is asked to address several questions:

- What would be the key scientific attributes of a predictive oversight system?
- What are the challenges in attempting to predict biological characteristics from sequence?
- Does the current state of the science of predicting function from sequence support a predictive oversight system at this time?
- If not, what are the scientific milestones that would need to be realized before a predictive oversight system might be feasible?
- In qualitative terms, what level of certainty would be needed about the ability to predict biological characteristics from sequence data in order to have confidence in a predictive oversight system?
- In what time frame might these milestones be realized? What kinds of studies are needed to achieve these milestones?

Appendix B

Committee Member and Staff Biographies

CHAIR

Dr. James W. LeDuc directs the Program on Global Health in the Institute for Human Infections and Immunity at the University of Texas Medical Branch. He also serves as deputy director of the Galveston National Laboratory. Previously, he served as the coordinator for influenza for the Centers for Disease Control and Prevention (CDC) in Atlanta, Georgia, and was the director of the Division of Viral and Rickettsial Diseases in the CDC National Center for Infectious Diseases (NCID). He began his professional career as a field biologist with the Smithsonian Institution's African Mammal Project in West Africa. He then served for 23 years as an officer in the U.S. Army Medical Research and Development Command. He joined CDC in 1992, was assigned to the World Health Organization as a Medical Officer, and later became the associate director for global health at NCID. His research interests include the epidemiology of arboviruses and viral hemorrhagic fevers and global health. He has participated in a number of National Research Council studies.

MEMBERS

Dr. Ralph Baric received his BS from North Carolina State University in 1977. He obtained his PhD from the Department of Microbiology of North Carolina State University in 1982, studying alpha-virus–host interaction and pathogenesis under the direction of Robert E. Johnston. He continued his postdoctoral training on coronavirus replication and pathogenesis under the direction of Michael M. C. Lai at the University of Southern California. In 1986, Dr. Baric was hired as an assistant professor in the Department of Parasitology and Laboratory Practice, and he is currently a professor in the Department of

Epidemiology and the Department of Microbiology and Immunology of the University of North Carolina at Chapel Hill. During his early training, Dr. Baric was a Harvey Weaver Scholar for the National Multiple Sclerosis Society and an established investigator for the American Heart Association in association with his studies of coronavirus replication, cross-species transmission, persistence, evolution, and pathogenesis. He is a member of the Editorial Board of the *Journal of Virology* and a senior editor for *PLoS Pathogens*. Dr. Baric is a permanent member of a National Institutes of health (NIH) study section (VirB); has been a consultant for the World Health Organization, the Centers for Disease Control and Prevention, and NIH; and has served on various institutional recombinant-DNA review committees. He has published over 130 peer-reviewed manuscripts, including several in *Science*, the *Proceedings of the National Academy of Sciences of the United States of America*, and *Nature Medicine*, and his research efforts are supported by several NIH research grants. Dr. Baric's expertise is primarily in norovirus molecular evolution and susceptibility and in coronavirus reverse genetics, synthetic genome reconstruction, pathogenesis, vaccine design, and cross-species transmission of viruses, often using the SARS coronavirus or noroviruses as models.

Dr. Roger G. Breeze received his veterinary degree in 1968 and his PhD in veterinary pathology in 1973, both from the University of Glasgow, Scotland. He was engaged in teaching, diagnostic pathology, and research on respiratory and cardiovascular diseases at the University of Glasgow Veterinary School from 1968 to 1977 and at Washington State University College of Veterinary Medicine from 1977 to 1987, where he was professor and chair of the Department of Microbiology and Pathology. From 1984 to 1987, he was deputy director of the Washington Technology Center, the state's high-technology sciences initiative, based in the College of Engineering of the University of Washington. In 1987, he was appointed director of the U.S. Department of Agriculture (USDA) Plum Island Animal Disease Center, a Biosafety Level 3 facility for research and diagnosis related to the world's most dangerous livestock diseases. In that role, he initiated research on the genomic and functional genomic basis of disease pathogenesis, diagnosis, and control of livestock RNA and DNA virus infections. That work became the basis of U.S. defense against natural and deliberate infection with those and led to his involvement in the early 1990s in biologic-weapons defense and proliferation prevention. From 1995 to 1998, Dr. Breeze directed research programs in 20 laboratories in the Southeast for the USDA Agricultural Research Service before going to Washington, D.C., to establish biologic-weapons defense research programs for USDA. He received the Distinguished Executive Award from President Clinton in 1998 for his work at Plum Island and in biodefense. Since 2004, he has been chief executive officer of Centaur Science Group, which provides consulting services in biodefense. His main commitment is to the Defense Threat Reduction Agency's

Biological Weapons Proliferation Prevention Program in Europe, the Caucasus, and Central Asia.

Dr. R. Mark Buller is widely recognized as a leader in the field of viral pathogenesis. His current research focuses on the interplay between the genetic expression of orthopoxviruses—such as monkeypox virus, ectromelia virus, and vaccinia virus—and the hosts' response to infection. Dr. Buller applies this work to the development of animal models for the evaluation of antivirals and vaccines for smallpox. He currently serves as a professor at Saint Louis University, Missouri, in the Department of Molecular Microbiology and Immunology. Dr. Buller is also director of the Aerosol Biology Core of the multi-institutional Midwest Regional Center for Excellence in Biodefense and Emerging Infectious Diseases Research. Before joining Saint Louis University, he was head of the Poxvirus Pathogenesis Group at the National Institute of Allergy and Infectious Diseases of the National Institutes of Health. Dr. Buller holds a PhD in virology from the Institute of Virology in Glasgow. He has published over 130 peer-reviewed scientific articles, reviews, and book chapters, and is a member of the editorial review boards of major scientific publications. Dr. Buller has also served as an invited reviewer, committee member, or speaker on the topic of bioterrorism and biomedical research.

Dr. Sean R. Eddy is a group leader at the Howard Hughes Medical Institute's Janelia Farm Research Campus outside Washington, D.C. His research interests are in the development of computational algorithms for genome-sequence analysis. He is the author of several widely used software tools for biologic sequence analysis, including a software package called HMMER; a coauthor of the Pfam database of protein domains; and a coauthor of the book *Biological Sequence Analysis: Probabilistic Models of Proteins and Nucleic Acids* (Cambridge University Press, 1998). He received a bachelor's degree from the California Institute of Technology and a PhD from the University of Colorado at Boulder, and he was a postdoctoral fellow at NeXagen Pharmaceuticals and at the MRC Laboratory of Molecular Biology. He was on the faculty of the Department of Genetics of the Washington University School of Medicine for 11 years before moving to Janelia Farm.

Dr. Stanley Falkow is the Robert W. and Vivian K. Cahill Professor of Microbiology and Immunology at Stanford University School of Medicine. He formulated molecular Koch's postulates, which have guided the study of the microbial determinants of infectious diseases since the late 1980s. Dr. Falkow received his BS from the University of Maine and went on to earn his PhD from Brown University. He discovered that infectious microorganisms use genes that are activated only inside host cells. Dr. Falkow has published numerous articles and has served on the editorial boards of several professional publications. In addition, he has received numerous awards for his achievements in science, including the Bristol-Myers Squibb Award for Distinguished Achievement in Infectious Disease Research, the Altemeier Medal from the Surgical Infectious Diseases Society of America, the Howard Taylor Ricketts Award Lecture at the University of Chicago, and the Paul Ehrlich–Ludwig Darmstaedter Prize. In 2003, he received the Abbott Lifetime Achievement Award from the American Society for Microbiology and the Selman A. Waksman Award in Microbiology from the National Academy of Sciences (NAS). He received the Robert Koch Award in 2000. Dr. Falkow was president of the American Society for Microbiology in 1997–1998. He was elected to the Institute of Medicine in 1997 and received the Maxwell-Finland Award from the National Foundation for Infectious Diseases in 1999. Also in 1999, he was named an honorary doctor of science by the University of Guelph, Canada, and received the University of Maine Alumni Career Award. He has received honorary doctorates in Europe and the United States. Dr. Falkow is a member of NAS and the National Academy of Arts and Sciences. He is also an elected fellow of the American Association for the Advancement of Science and a foreign member of the UK Royal Society. Dr. Falkow was nominated twice for a Nobel Prize in physiology or medicine. In 2008, Dr. Falkow received the Lasker Award for medical research.

Ms. Rachel E. Levinson, a 25-year veteran of science policy at the national level, is the director of the Arizona State University (ASU) Washington office and is responsible for special projects and research initiatives in the Office of the Vice President for Research and Economic Affairs. She joined ASU in 2005 as the director of the Government and Industry Liaison Office, Biodesign Institute at Arizona State University. Ms. Levinson heads an office responsible for facilitating mutually beneficial relationships between university researchers, federal funding agencies, and private-sector entities. Most recently, she was with the Office of Science and Technology Policy in the Executive Office of the President, where she was assistant director of life sciences. She began her career as a biologist at the National Cancer Institute of the National Institutes of Health (NIH). She advanced to become deputy director of the NIH Office of Recombinant DNA and senior policy adviser in the Office of Technology Transfer. Ms. Levinson earned her BS in zoology from the University of Maryland at College Park and her MA in science, technology, and public policy from the George Washington University School of Public and International Affairs.

Dr. John Mulligan founded Blue Heron Biotechnology in 1999 after a decade of genomics research, including establishment and management of one of the two Human Genome Centers at Stanford University and direction of genomics research at Darwin Molecular. Blue Heron Biotechnology is a pioneer in and leader of the gene-synthesis market.

Dr. Alison D. O'Brien is a professor in and chair of the Department of Microbiology and Immunology at the Uniformed Services University of the Health Sciences. She is a past president of the American Society for Microbiology. She received her PhD from Ohio State University in 1976. Research in Dr. O'Brien's laboratory focuses on the molecular mechanisms by which the Shiga toxins from enterohemorrhagic *Escherichia coli* (EHEC) contribute to hemorrhagic colitis and the hemolytic uremic syndrome, the involvement of toxins from uropathogenic *E. coli* (UPEC) in the host response to urinary tract infections, and development of therapeutics against infections caused by *Bacillus anthracis* and *B. cereus*.

Dr. Francisco Ochoa-Corona, a forensic plant pathologist, specializes in delivering and developing reference diagnostics for exotic, naturalized, and indigenous plant viruses and other phytopathogens of relevance to agricultural biosecurity. His work is applicable to plant pathogens that can be intercepted at the border or detected through general surveillance in field settings or in transitional facilities. Dr. Ochoa-Corona's research in plant pathology contributes scientific input to regulatory officials regarding plant health emergencies. He joined Oklahoma State University in 2008 from the Investigation and Diagnostic Centre of Biosecurity New Zealand, in the Ministry of Agriculture and Forestry, where he was principal adviser in virology.

Prof. Jane S. Richardson earned a bachelor's degree from Swarthmore College and a master's degree from Harvard University in 1966. Since 1970, she has been at Duke University Medical Center, where she and her husband, David, work together in investigating the three-dimensional structure of proteins and RNA. They were early pioneers in protein crystallography, in molecular computer graphics, and in the field of de novo protein design, proposing and then making novel amino-acid sequences designed to fold into specific 3D structures. Prof. Richardson was the developer of ribbon drawings of protein structures, originally done by hand but since universally adopted in molecular graphics. She identified many well-known structural motifs, such as the helix N-cap, and has recently concentrated on new methods for the validation and improvement of protein and RNA crystal structures. She became a MacArthur Fellow in 1985, a member of the National Academy of Sciences in 1991, and a member of the Institute of Medicine in 2006, and she was an assessor for last year's CASP8 structure predictions.

Dr. Margaret Riley is a professor of biology at the University of Massachusetts Amherst (Umass Amherst). She received her PhD in population genetics from Harvard University and performed postdoctoral research in microbial population genetics with a Sloan Postdoctoral Fellowship in Molecular Evolution. She joined the faculty of Yale in 1991 and recently moved to UMass Amherst. She has a broad set of research interests that range from studies of experimental evolution of microorganisms to development of novel antimicrobials and redefinition of the microbial species concept. Dr. Riley studies the evolution of microbial diversity with an emphasis on the ecology and evolution of microbial toxins. Her recent work has revealed that the production of toxins is a primary force in the generation and maintenance of microbial diversity. Those studies led to an interest in applying ecologic and evolutionary theory to the design of novel antimicrobials for use in animal and human health. She is cofounder of Origin Antimicrobials, Inc., whose mission is to discover and refine novel antimicrobials to address the challenge of antibiotic resistance. Dr. Riley is the director of the Organismic and Evolutionary Biology Program and the director of the Museum of Natural History at UMass Amherst. From 1999 to 2002, she chaired the Gordon conference on molecular evolution; from 2003 to 2005, she chaired the Gordon conference on microbial population biology and evolution. She is a fellow of the American Academy of Microbiology.

Mr. Tom Slezak has been involved with bioinformatics at Lawrence Livermore National Laboratory (LLNL) for 30 years after receiving a BS and an MS in computer science from the University of California, Davis. He is currently the associate program leader for informatics for the Global Security Program efforts at LLNL. He was involved in the Human Genome Program from 1987 to 2000, leading the informatics efforts at LLNL and then the Department of Energy Joint Genome Institute from 1997 to 2000. In 2000, he began to build a pathogen bioinformatics team at LLNL, pioneering a novel whole-genome analysis approach to DNA signature design. His team developed signature targets for multiple human pathogens that were used at the 2002 Winter Olympic Games under the BASIS program and later adapted for use nationwide in the Department of Homeland Security (DHS) BioWatch program. Under a close collaboration with the Centers for Disease Control and Prevention, the LLNL team has been called on for computational help on smallpox, SARS, monkeypox, avian influenza, and numerous other diseases. In addition to continuing work on human and agricultural pathogens, Mr. Slezak team is focusing on signatures of mechanisms of virulence, antibiotic resistance, and evidence of genetic engineering. They have been working on detecting novel, engineered, and advanced biothreats for several years, leveraging high-risk Information Technology Industry Council and DHS funding. Mr. Slezak has chaired or served on multiple advisory boards, including those of the rice genome project, mouse and maize genetics databases, the spruce tree genome project (Canada), plant pathogens, and a National Institute of Allergy and Infectious Diseases sequencing center contract renewal.

NATIONAL RESEARCH COUNCIL STAFF

Dr. India Hook-Barnard is a program officer with the Board on Life Sciences of the National Research Council. She came to the National Academies from the National Institutes of Health where she was a Postdoctoral Research Fellow from 2003 to 2008. Her research investigating the molecular mechanism of gene expression focused on the interactions between RNA polymerase and promoter DNA. Dr. Hook-Barnard earned her PhD from the Dept. of Molecular Microbiology and Immunology at the University of Missouri. Her graduate research examined translational regulation and ribosome binding in *Escherichia coli*. At the National Academies, she contributes to projects in a variety of exciting topic areas. Much of her current work is related to issues of biosecurity, microbiology, and genomics. She is the study director or staff officer for several ongoing projects including the U.S. Canada Regional Committee for the International Brain Research Organization, Animal Models for Assessing Countermeasures to Bioterrorism Agents, and Framework for Developing a New Taxonomy of Disease.

Mr. Carl-Gustav Anderson is a senior program assistant with the Board on Life Sciences of the National Research Council. He received a BA in philosophy from American University in 2009, completing significant research projects exploring on the philosophy of the Kyoto School. He has worked closely with the All Women's Action Society (Malaysia), helping to engage young men in feminist dialogue and to present a feminist response to the unique identity politics of contemporary Malaysia. He has focused his research interests on Buddhist encounters with the West, with particular emphasis on Buddhist responses to Western feminism, communism, transcendental philosophy, and existentialism.

Since joining the Board on Life Sciences in 2009, he has served as senior program assistant for *Responsible Research with Biological Select Agents and Toxins* (2009) and *Challenges and Opportunities for Education about Dual Use Issues in Life Sciences Research* (2010). In addition to several consensus committees, he also serves as senior program assistant for the United States-Canada Regional Committee to the International Brain Research Organization.

Appendix C

HHS and USDA Select Agents and Toxins

HHS AND USDA SELECT AGENTS AND TOXINS
7 CFR Part 331, 9 CFR Part 121, and 42 CFR Part 73

HHS SELECT AGENTS AND TOXINS

Abrin
Botulinum neurotoxins
Botulinum neurotoxin producing species of *Clostridium*
Cercopithecine herpesvirus 1 (Herpes B virus)
Clostridium perfringens epsilon toxin
Coccidioides posadasii/Coccidioides immitis
Conotoxins
Coxiella burnetii
Crimean-Congo haemorrhagic fever virus
Diacetoxyscirpenol
Eastern Equine Encephalitis virus
Ebola virus
Francisella tularensis
Lassa fever virus
Marburg virus
Monkeypox virus
Reconstructed replication competent forms of the 1918 pandemic influenza virus containing any portion of the coding regions of all eight gene segments (Reconstructed 1918 Influenza virus)
Ricin
Rickettsia prowazekii
Rickettsia rickettsii
Saxitoxin
Shiga-like ribosome inactivating proteins
Shigatoxin
South American Haemorrhagic Fever viruses
  Flexal
  Guanarito
  Junin
  Machupo
  Sabia
Staphylococcal enterotoxins
T-2 toxin
Tetrodotoxin
Tick-borne encephalitis complex (flavi) viruses
  Central European Tick-borne encephalitis
  Far Eastern Tick-borne encephalitis
  Kyasanur Forest disease
  Omsk Hemorrhagic Fever
  Russian Spring and Summer encephalitis
Variola major virus (Smallpox virus)
Variola minor virus (Alastrim)
Yersinia pestis

OVERLAP SELECT AGENTS AND TOXINS

Bacillus anthracis
Brucella abortus
Brucella melitensis
Brucella suis
Burkholderia mallei (formerly *Pseudomonas mallei*)
Burkholderia pseudomallei (formerly *Pseudomonas pseudomallei*)
Hendra virus
Nipah virus
Rift Valley fever virus
Venezuelan Equine Encephalitis virus

USDA SELECT AGENTS AND TOXINS

African horse sickness virus
African swine fever virus
Akabane virus
Avian influenza virus (highly pathogenic)
Bluetongue virus (exotic)
Bovine spongiform encephalopathy agent
Camel pox virus
Classical swine fever virus
Ehrlichia ruminantium (Heartwater)
Foot-and-mouth disease virus
Goat pox virus
Japanese encephalitis virus
Lumpy skin disease virus
Malignant catarrhal fever virus
  (Alcelaphine herpesvirus type 1)
Menangle virus
Mycoplasma capricolum subspecies *capripneumoniae* (contagious caprine pleuropneumonia)
Mycoplasma mycoides subspecies *mycoides* small colony (*Mmm*SC) (contagious bovine pleuropneumonia)
Peste des petits ruminants virus
Rinderpest virus
Sheep pox virus
Swine vesicular disease virus
Vesicular stomatitis virus (exotic): Indiana subtypes VSV-IN2, VSV-IN3
Virulent Newcastle disease virus[1]

USDA PLANT PROTECTION AND QUARANTINE (PPQ) SELECT AGENTS AND TOXINS

Peronosclerospora philippinensis (*Peronosclerospora sacchari*)
Phoma glycinicola (formerly *Pyrenochaeta glycines*)
Ralstonia solanacearum race 3, biovar 2
Rathayibacter toxicus
Sclerophthora rayssiae var zeae
Synchytrium endobioticum
Xanthomonas oryzae
Xylella fastidiosa (citrus variegated chlorosis strain)

11/17/2008

[1] A virulent Newcastle disease virus (avian paramyxovirus serotype 1) has an intracerebral pathogenicity index in day-old chicks (Gallus gallus) of 0.7 or greater or has an amino acid sequence at the fusion (F) protein cleavage site that is consistent with virulent strains of Newcastle disease virus. A failure to detect a cleavage site that is consistent with virulent strains does not confirm the absence of a virulent virus.

Appendix D

2009 Workshop Agenda

Scientific Milestones for the Development of a Gene-Sequence-Based Classification System for Oversight of Select Agents

Thursday, Sept. 3rd, 2009
The National Academy of Sciences Building: Lecture Room
2100 C St., N.W. • Washington, D.C. 20037

AGENDA

8:30 a.m. **Welcome and Opening Remarks**

James LeDuc, committee chair—*The University of Texas Medical School*

The workshop in context of the study and the statement of task

9:00 a.m. **Session 1: The Current Structure for Oversight**

What are the current forms of oversight? Are there gaps in the oversight, and if so, are these gaps emerging as a result of new technology, new user communities, or new perceptions? How might a sequence based system be helpful in addressing these gaps/ concerns?

**Moderator: Rachel Levinson*

- **Julia Kiehlbauch**, *United States Department of Agriculture, Animal and Plant Health Inspection Service*

- **Rob Weyant**, *Centers for Disease Control and Prevention*—Synthetic DNA and the Select Agent Regulations.
- **Claudia Mickelson**, *Massachusetts Institute of Technology*—IBC, RAC guidelines and concerns about sequences.
- **Edward You**, *Federal Bureau of Investigation*—Surveillance of Select Agent list and emerging concerns.
- **Amy Patterson**, *National Institutes of Health, Office of Biotechnology Activities*—Comprehensive view and the need for this study.

Panel discussion: ~30 min

10:30 a.m. **Break**

10:45 a.m. **Session 2: Current Mechanisms and Criteria for Screening and Surveillance**

What is currently being done? How are sequences chosen to monitor? What is a "sequence of concern"?

**Moderator—John Mulligan*

- **Pete Pesenti**, *Department of Homeland Security*—What are the factors and process used for risk assessment? What are the criteria or characteristics of agents (or sequences) considered a threat?
- **John Mulligan**, *Blue Heron Biotechnology*—What are the current screening practices, standards, and procedures in the industry? What are challenges and concerns?
- **Marcus Graf**, *GeneArt* and **Claes Gustafsson**, *DNA 2.0*—Representing companies working to harmonize screening techniques. What would they like to know to help the decision making process?
- **Stephen M. Maurer**, *University of California at Berkeley*—Interface of biosecurity, synthetic biology, and industry.

Panel discussion: ~30 min ** ***Ed You, FBI will join panel*** **

12:15 p.m. **Lunch**

1:00 p.m. **Session 3: Virulence**

What is virulence? Why is it so hard to predict? What attributes make a pathogen a threat to biosecurity? —to public health? Is there a difference?

**Moderator—Stan Falkow*

- **Stan Falkow**, *Stanford University*—Overview of virulence, meaning of genomics in prediction.
- **Jeff Taubenberger**, *National Institutes of Health, National Institute of Allergy and Infectious Diseases*—Influenza virulence and the role of genotype-phenotype relationships.
- **Michael Katze**, *University of Washington*—Systems biology and the difficulty predicting the importance of a sequence.
- **Ralph Baric**, *University of North Carolina at Chapel Hill*—SARS, systems genetics and pathogenesis.
- **Ramon Felciano**, *Ingenuity Systems*—Systems biology and pathway modeling of pathogenesis and host response.

Panel discussion: ~30 min

3:10 p.m. **Break**

3:25 p.m. **Session 4: Predicting Pathogenicity from Sequence**

Speakers will address gaps, challenges, and timeframe for milestones.

**Moderator—Sean Eddy*

- **Sean Eddy**, *Howard Hughes Medical Institute, Janelia Farm Research Campus*—Overview of sequence analysis; how reliably can protein function be predicted from protein sequence?
- **Jonathon Eisen**, *University of California at Davis*—Phylogenomic inference of protein function and the importance of genomic context.

- **Elliot J Lefkowitz**, *University of Alabama at Birmingham*—Bioinformatics support for pathogen research; Viral gene discovery and pathogenic potential.
- **John Moult**, *University of Maryland Biotechnology Institute, Center for Advanced Research in Biotechnology*—Protein structure and function prediction.
- **Ian Lipkin**, *Columbia University, Mailman School of Public Health*—Identification of emerging or novel microorganisms—pathogen surveillance.

Panel discussion: ~30 min

5:45 p.m. Closing Remarks

6:00 p.m. Adjourn

Appendix E

Applicability of the Select Agent Regulations to Issues of Synthetic Genomics

In a December 2006 report entitled "Addressing Biosecurity Concerns Related to the Synthesis of Select Agents" (www.biosecurityboard.gov/links.asp), the National Science Advisory Board on Biosecurity recommended that the federal government take steps to "Increase awareness among providers and users of synthetic genomic materials regarding compliance with the select regulations; and provide a list of genomic materials explicitly covered by the regulations."

The purpose of this document is to provide guidance regarding the application of the current select agent regulations to those who create and use synthetic genomic products. The current select agent regulations implement the provisions of the Agricultural Bioterrorism Act of 2002 and the Public Health Security and Bioterrorism Preparedness and Response Act of 2002. Select agents are bacteria, viruses, fungi, other microorganisms and toxins that have been deemed to have the potential to pose a significant risk to public health, plant or animal health, or plant or animal production. Regulation of the possession, use, and transfer of select agents is implemented by the U.S. Department of Agriculture Animal and Plant Health Inspection Service (USDA/APHIS) and the U.S. Department of Health and Human Services, Centers for Disease Control and Prevention (HHS/CDC). Individuals applying for access to select agents must undergo a security risk assessment by the Federal Bureau of Investigation, Criminal Justice Information Service (FBI/CJIS). Information on the select agent regulations (42 CFR Part 73, 9 CFR Part 121, and 7 CFR Part 331) can be found at the national select agent Web site (www.selectagents.gov).

The select agent regulations provide that the following genetic elements, recombinant nucleic acids, and recombinant organisms are select agents (See section 3(c) of 42 CFR Part 73, 9 CFR Part 121, and 7 CFR Part 331):

- Nucleic acids that can produce infectious forms of any of the select agent viruses.
- Recombinant nucleic acids that encode for the functional form(s) of select agent toxins if the nucleic acids:
  - Can be expressed *in vivo* or *in vitro* or,
  - Are in a vector or recombinant host genome and can be expressed *in vivo* or *in vitro*.
- Select agents and toxins that have been genetically modified.

The purpose of this regulatory language is to address advancements in molecular biology that may influence the production of infectious forms of select agent viruses, or the active forms of select agent toxins. It has been demonstrated, for example, that the single stranded (positive strand) RNA viruses and certain double stranded DNA viruses that utilize host polymerases contain nucleic acids that can produce infectious forms. Examples of select agent viruses that meet this criterion, and would therefore be regulated, include:

- Tickborne encephalitis complex (flavi) viruses:
  - Central European Tick borne encephalitis
  - Far Eastern Tickborne encephalitis
  - Russian Spring and Summer encephalitis
  - Kyasanur Forest Disease
  - Omsk Hemorrhagic Fever
- Eastern Equine Encephalitis virus
- Venezuelan Equine Encephalitis virus
- Classical Swine Fever Virus
- Foot-And-Mouth Disease Virus
- Japanese Encephalitis Virus
- Swine Vesicular Disease Virus
- Cercopithecine Herpesvirus 1 (Herpes B virus)
- Malignant Catarrhal fever Virus (Alcelaphine Herpesviurs Type 1)

Under the current select agent regulations the following are examples of materials that would not be regulated as a select agent:

- Non-infectious components of select agent viruses including:
  - Material from regulated genomes that has been rendered non-infectious
  - cDNA made from regulated select agent genomes
  - Genomic fragments from select agents (unless they encode for a functional form of a select agent toxin)
  - Complete genomes of single-stranded negative-strand RNA viruses, double-stranded RNA viruses, and double-stranded DNA

viruses that require a unique polymerase (Variola major virus,* Monkeypox virus, African swine fever virus, camelpox virus, Goat pox virus, lumpy skin disease virus, and Sheep pox virus)
  - Genomic material from select agent bacteria or fungi

(*It should be noted that, although the current select agent regulations to not apply to Variola major genetic elements, the World Health Organization places significant restrictions on the possession, use, and transfer of these materials. Institutions other than the two currently recognized WHO collaborating centers may not possess genetic fragments exceeding 20 percent of the Variola virus genome. For additional information on WHO Guidelines for Variola virus research, please see http://www.who.int/csr/disease/smallpox/research/en/index.html, and the report, also published in the Weekly Epidemiologic Record in 2008, on permissible use of variola genetic material http://www.who.int/csr/disease/smallpox/SummaryrecommendationsMay08. pdf).

- Genomic material from Select Agent strains that have been excluded from regulation under section 3(e) of the select agent regulations

Additionally, select agent nucleic acid sequence information is not regulated.

Individuals or entities that possess, use, or transfer select agents must meet all of the requirements of the select agent regulations (42 CFR Part 73, 7 CFR Part 331, and 9 CFR Part 121) prior to possession, use, or transfer. These regulations, the associated enabling legislation, related guidance documents, registration forms, and contact information for the Select Agent Programs can be found at the National Select Agent Registry website (www.selectagents.gov).

The following examples, while not inclusive of all potential scenarios, illustrate the application of the current select agent regulations to activities involving synthetic genomics or synthetic biology.

Example scenarios involving synthetic genomic select agent materials.

1. An individual submits to a producer (facility that manufactures the material) a full genome sequence of Foot-and-Mouth Disease Virus (FMDV) and requests that it be synthesized and shipped to the submitter (individual that requests the material).

   - Does the processing of this order fall under the current select agent regulations?

Yes. An individual or entity in possession of the full FMDV genome, regardless of how the individual or entity came into possession of it, or for however brief a time period, would be required to be preregistered under provisions of the select agent regulations, meeting all of the safety, security and personnel reliability requirements therein. Transfer of the full FMDV genome would require prior approval from the select agent program.

- Does the sequence information that this individual submitted fall under the current select agent regulations?

No. The current regulations do not cover sequence information.

- The producer is registered with the Select Agent Program for possession of infectious FMDV genomic material. Once this material is produced, can the producer send it to the submitter?

Yes, with stipulations. First, the submitter must also be registered with the Select Agent Program to possess select agents. Second, all domestic transfers of select agents must be preauthorized by either the CDC or APHIS Select Agent Program. All international exports must be preauthorized by the Department of Commerce. Since the producer is registered with the Select Agent Program, the program must first authorize the domestic transfer. Instructions and forms for use in obtaining this authorization are available on the National Select Agent Registry Web site (www.selectagents.gov).

2. An individual submits an unidentified sequence to a producer and asks for its synthesis. Screening of this sequence by the producer shows a high degree of homology with the pXO2 virulence-associated plasmid of *Bacillus anthracis*.

- Does the processing of this order fall under the current Select Agent Regulations?

No. Although *B. anthracis* organisms are regulated, the current regulations do not cover individual *B. anthracis* genetic elements. Unless the functional form of a select agent toxin is included in the product, select agent bacterial genomic material is not covered by the current regulations. However, all international exports of *B. anthracis* genetic elements associated with pathogenicity must be preauthorized by the Department of Commerce. Instructions for obtaining an export license are available on the Bureau of Industry and Security Web site, www.bis.doc.gov.

3. An individual submits a genomic sequence for *Y. pestis* and asks for its synthesis.

   - Does the processing of this order fall under the current Select Agent Regulations?

   No. Unless the functional form of a select agent toxin is included in the product, select agent bacterial and fungal genomes are not covered under the regulations. However, all international exports of *Y. pestis* genetic elements associated with pathogenicity must be preauthorized by the Department of Commerce. Instructions for obtaining an export license are available on the Bureau of Industry and Security Web site, www.bis.doc.gov.

4. An investigator in Canada submits an order for the synthesis and delivery to Canada of the genome of the Omsk Hemorrhagic Fever virus from a producer located in the United States.

   - Does the processing of this order fall under the current Select Agent Regulations?

   Yes. The possession of this material is regulated by the Select Agent Regulations and the entity performing the synthesis would have to be registered with the Select Agent Program. However, because the material is to be exported, it will require an export license from the U.S. Department of Commerce, Bureau of Industry and Security instead of a Select Agent Program authorization. Instructions for obtaining this license are available on the Bureau of Industry and Security Web site, www.bis.doc.gov.

5. A foreign national works at a laboratory and is trained to grow Eastern Equine Encephalitis virus (EEE) in order to produce a vaccine.

   - Does this activity fall under the current Select Agent Regulations?

   Yes. The facility housing this work would have to be registered with the Select Agent Program and the foreign national involved in this work would have to have a clear Select Agent Security Risk Assessment in order to access the virus. In addition, there are "deemed export" licensing requirements from the Department of Commerce for the transfer of production technology to the foreign national working in the United States. An export license is required to transfer the technology to produce the EEE (1E001). Even though the final end use is a vaccine which is controlled under 1C991, the technology involves the virus strain and thus controlled. Instructions for obtaining an export license are available on the Bureau of Industry and Security Web site, www.bis.doc.gov.

6. A synthetic genomics producer located outside the U.S. receives an order from a laboratory within the U.S. for the genome of the Russian Spring and Summer encephalitis virus.

   - Does this activity fall under the current Select Agent Regulations?

   Yes, in part. Although the producer is not required to follow the Select Agent Regulations, the receiving laboratory must be registered with the Select Agent Program. In addition, the receiving laboratory must obtain an import permit from the CDC's Etiologic Agent Import Permit Program prior to importing this material. Information on obtaining import permits can be obtained from the National Select Agent Registry Web site (www.selectagents.gov).

Appendix F

Summary of Relevant Legislation, Regulation, and Guidance

- **Biological and Toxin Weapons Convention (BWC):** Signed in 1972 and entered into force in 1975,[1] the BWC functions as an international effort to control biological weapons and focuses on the prohibition of the development, production, and stockpiling of biological and toxin weapons. Of most immediate relevance is Article I, which states: "Each State Party to this Convention undertakes never in any circumstances to develop, produce, stockpile or otherwise acquire or retain: (1) Microbial or other biological agents, or toxins whatever their origin or method of production, of types and in quantities that have no justification for prophylactic, protective or other peaceful purposes; (2) Weapons, equipment or means of delivery designed to use such agents or toxins for hostile purposes or in armed conflict.It is important to note that the BWC neither prohibits research on defenses against biological weapons, nor establishes any list of prohibited agents."

- **The Biological Weapons Anti-Terrorism Act of 1989** (Public Law 101-298, May 22, 1990): To implement the norms established by the BWC, the United States enacted the BWATA, which "established penalties for violating the Convention's prohibitions, unless "(1) such biological agent, toxin, or delivery system is for a prophylactic, protective, or other peaceful purpose; and (2) such biological agent, toxin, or delivery system, is of a type and quantity reasonable for that purpose." In keeping with the treaty, the legislation focused on the *purpose* for which agents or toxins

[1]The BWC treaty's formal title is the Convention on the Prohibition of the Development, Production and Stockpiling of Bacteriological (Biological) and Toxin Weapons and on Their Destruction.

were possessed, rather than the agents themselves. The law authorizes the government to apply for a warrant to seize any biological agent, toxin, or delivery system that has no apparent justification for peaceful purposes, but prosecution under the law would require the government to prove that an individual did not have peaceful intentions (Atlas 1999).(BSAT 33)" Critically, the BWATA reified the tenets of the BWC by establishing the first explicit criminal punishments for the development, manufacture, transfer, or possession of a biological agent, toxin, or delivery system for use as a weapon. The BWC and its implementation in the form of the BWATA form the groundwork of the Select Agent Regulations.

- **The Antiterrorism and Effective Death Penalty Act of 1996** (Public Law 104-132, April 24, 1996) provides the first instance of list-based attempts at the regulation of biological agents. "The Act required the Secretary of Health and Human Services (HHS) to issue regulations to govern the transport of biological agents with the potential to pose a severe threat to public health and safety through their use in bioterrorism. In establishing the list of materials to regulate, the Secretary was to consider: '(I) the effect on human health of exposure to the agent; (II) the degree of contagiousness of the agent and the methods by which the agent is transferred to humans; (III) the availability and effectiveness of immunizations to prevent and treatments for any illness resulting from infection by the agent; and (IV) any other criteria that the Secretary considers appropriate' (Public Law 104-132, April 24, 1996, Sec. 511). The Secretary delegated the authority to regulate these "select agents" to the Centers for Disease Control and Prevention (CDC). To ensure that the transfer of these agents was carried out only by and between responsible parties, the CDC required that laboratories transferring select agents be registered and report each transfer"[2] (NRC 2009b).

- **Uniting and Strengthening America by Providing Appropriate Tools Required to Intercept and Obstruct Terrorism Act of 2001** (Public Law Pub. L. 107-56) expands on the BWATA, making "it an offense for a person to

[2] "The purpose of registration was to control domestic transfers based upon a permitting system. A registered laboratory could legally transfer select agents only to another registered laboratory; some transfers were denied because of concerns about the adequacy of the facility proposed to receive the agent. Transfers to nonregistered laboratories were prohibited. Registration, however, was principally a matter of notification: A laboratory was obligated to notify relevant authorities of a transfer to another registered facility and that the transfer itself complied with applicable safety standards. Specific information about particular pathogens that the facility possessed did not have to be reported, not even if they were the subjects of extensive research, so long as they were not transferred. This was not intended to be a strict licensing system but merely a way of overseeing transfers and shipments of lethal pathogens" (NRC 2004).

knowingly possess any biological agent, toxin, or delivery system of a type or in a quantity that, under the circumstances, is not reasonably justified by prophylactic, protective, bona fide research, or other peaceful purpose" (NRC 2009b).

- **Public Health Security and Bioterrorism Preparedness and Response Act**, known as the Bioterrorism Act of 2002 (Public Law 107–188, June 12, 2002): "This Act added requirements for regulations governing possession of select agents, including approval for laboratory personnel by the Attorney General following a background check by the Federal Bureau of Investigation (FBI). It also gave the U.S. Department of Agriculture (USDA), through its Animal and Plant Health Inspection Service (APHIS), the authority to regulate the possession, use, and transfer of BSAT that relate to plant and animal health and products, complementing the authority granted to CDC for human pathogens. The regulation of select agents and toxins is thus a shared federal responsibility involving HHS/CDC, USDA/APHIS, and the Department of Justice (DOJ). The Bioterrorism Act has been implemented through a series of regulations; the final regulations—42 CFR 73 (human pathogens), 9 CFR 121 (animal pathogens), and 7 CFR 331 (plant pathogens)—became effective in the spring of 2005.[3] (NRC 2009)"

- **Agricultural Bioterrorism Protection Act of 2002; Possession, Use, and Transfer of Biological Agents and Toxins (7 CFR 331; 9 CFR 121)** established the initial list of biological agents and toxins determined to have the potential to pose a severe threat to animal or plant health, or to animal or plant products.

- **Applicability of the Select Agent Regulations to Issues of Synthetic Genomics:** The CDC has provided guidance regarding the application of the current Select Agent Regulations to those who create and use synthetic genomic products. Specifically, the guidance defines the organisms whose genomes are covered and provides examples that clarify the application of these rules (Appendix E).

- **Screening Framework Guidance for Synthetic Double-Stranded DNA Providers:** HHS has issued guidance to Synthetic DNA Providers, which requests that providers identify and follow up on sequences with homology *unique* to a Select Agent sequence. Thus, the guidance aims to define the boundaries between the sequence of a a Select Agent and a similar sequence from a related species.

[3]Agents that can affect both human and animals, called "overlap agents," are listed in both the CDC and USDA lists.

- **Executive Order July 2, 2010: Optimizing the Security of Biological Select Agents and Toxins in the United States** directs federal agencies to institute changes in the current implementation of the Select Agent Regulations. The Executive Order also directs federal agencies to take actions to improve the overall coordination, consolidation and oversight of select agents and toxins: (See Appendix M)

Appendix G

Influenza A and SARS-CoV

INFLUENZA A

The influenza A virus proapoptotic PB1-F2 protein has been clearly implicated as a major virulence factor in some highly pathogenic influenza virus strains, but the H1N1 swine-origin influenza virus pandemic strain codes for a truncated PB1-F2 protein that terminates after 11 amino acids. It was predicted that the truncation would attenuate swine influenza pathogenesis, identifying a possible key mutation that could emerge to enhance H1N1 virulence and contribute to an expanding epidemic. However, disruption of PB1-F2 expression in several other influenza virus backgrounds or by intermixing functional PB1-F2 between strains had little effect on viral lung load in mice. The data suggests that the PB1-F2 virulence determinant may be context- or host-dependent, perhaps by enhancing virulence by other mechanisms that are independent of replication. The effects of restoring a full length functional PB1-F2 protein on 2009 swine H1N1 in vivo pathogenesis are difficult to predict because its virulence-enhancing activities may depend on co-evolutionary changes elsewhere in the genome. As a working model for predicting virulence from sequence information, the preponderance of influenza data suggest that restoration of a full length PB1-F2 protein will enhance the virulence of swine H1N1—a hypothesis that will probably be tested using reverse genetics in the near future (McAuley, Zhang et al.)

SARS-COV

It is also clear that distantly related viral proteins can interact with a conserved cellular protein target and thereby augment their pathogenic potential. Among coronaviruses as with many other viruses, receptor interactions are an important determinant of species specificity, tissue tropism, virulence, and pathogenesis. Pathogenesis depends upon the ability of a virus to dock and enter into a suitable human host cell. For example, the highly pathogenic emerging group 2 coronavirus that causes severe acute respiratory syndrome, coronavirus (SARS-CoV) and a distantly related less pathogenic group 1 human coronavirus, NL63-CoV, both encode a large 180/90kDa spike glycoprotein (S) that engages a host cellular receptor(s) to mediate docking and entry into cells. The SARS-CoV and NL63-CoV S glycoproteins are about 40 percent identical and encode novel, yet unrelated receptor binding domains (RBD) in S that engage the same cellular receptor, angiotensin-converting enzyme 2 (ACE2) to mediate virus docking and entry into cells. Despite the absence of structural homology in the RBD cores of NL63-CoV and SARS-CoV, the two viruses recognize common ACE2 regions by using novel protein-protein folds and interaction networks. On the basis of sequence, it was not possible to predict that the two highly divergent coronavirus RBDs would engage a similar "hot spot" on the surface of the ACE2 receptor and thus mediate docking and entry into cells. Moreover, the pathogenic potential of the two human coronaviruses are distinct: SARS-CoV causes an atypical pneumonia that results in acute respiratory distress syndrome with mortality exceeding 50 percent in people over 60 years old, whereas NL63-CoV causes a self-limiting denuding bronchiolitis and croup, primarily in infants and children. Clearly, other factors besides virus-receptor interaction and entry are regulating severe acute end-stage lung-disease outcomes during SARS-CoV infection, and this complicates sequence-based predictions of virus-receptor interaction networks and virulence outcomes (Proc Natl Acad Sci U S A. 2009 Nov 24;106(47):19970-19974. Epub 2009 Nov 9. Crystal structure of NL63 respiratory coronavirus receptor-binding domain complexed with its human receptor (Wu, Li et al. 2009).

Appendix H

Virus-Host Interactions

Virus-host interactions play a critical role in regulating disease severity and distribution in human populations. Moreover, it is likely that pathogenic microorganisms have shaped the genetic population structure of humans. For example, noroviruses are category B biodefense pathogens and the primary etiologic agent responsible for epidemic viral gastroenteritis worldwide. Members of this diverse family of viruses are the most common causes of sporadic diarrhea in community settings and a major burden on the military, restaurant services, the cruise-ship industry, university campuses, hospitals, and retirement communities. Humans encode a highly diverse set of histo-blood group (HBGA) carbohydrates on mucosal surfaces that are regulated by several highly polymorphic fucosyltransferase genes designated FUT1, FUT2, and FUT3 and by the enzymes that regulate A and B carbohydrate expression, resulting in dramatic differences in HBGA expression in human populations. Several studies have indicated that different HBGAs function as the receptors or coreceptors for productive norovirus infection in humans. People who cannot express HBGAs on mucosal surfaces ($FUT2^{-/-}$) are highly resistant to Norwalk virus (NV) and perhaps other noroviruses, whereas people who express O type HBGAs on mucosal surfaces are more susceptible to NV. The most prevalent strains (GII.4) caused global pandemics of severe gastroenteritis in 1996, 2002, and 2006. Epidemic GII.4 viruses appear to have evolved two techniques to maintain their high prevalence in human populations. First, new epidemic GII.4 variants have emerged from ancestral strains and have altered HBGA-receptor binding profiles, allowing new strains to target unique susceptible human population groups that were probably resistant to ancestral strains. Second, like influenza viruses, exigent GII.4 norovirus variants undergo antigenic variation and so escape herd immunity. Thus, it is clear that host genetics have profound influences in regulating susceptibility to and virulence of *viruses* (Norovirus pathogenesis: mechanisms of persistence and immune evasion in human populations. Donaldson EF, Lindesmith LC, Lobue AD, Baric RS. Immunol Rev. 2008 Oct;225:190-211; Human susceptibility and resistance to Norwalk virus infection (Lindesmith, Moe et al. 2003).

Appendix I

Botulinum Neurotoxin, *B. Anthracis* and Variola Virus

This appendix describes the complexity of the pathogenic mechanisms of the select agents botulinum neurotoxin, *Bacillus anthracis*, variola virus, filoviruses, and coronaviruses. It discusses the general mechanisms for acquisition of pathogenicity and the important role of the host in the calibration of the pathogenic potential of a microorganism

Botulinum neurotoxin (BoNT), *B. anthracis*, and variola virus have a majority of the attributes of pathogenicity that are considered important for inclusion on the Select Agents list. The toxin and both pathogens were featured in at least one nation-state bioweapons program in the last century, are associated with high case-fatality rates, and are relatively easy to produce or grow. However, there are important differences among them. BoNT is easy to isolate from *Clostridium botulinum*, is the most poisonous substance known, but is difficult to disseminate. *B. anthracis* is found in nature in the United States, is extremely stable in the environment, and is also poorly transmissible from human to human. Variola virus no longer circulates in the human population, is less stable in the environment, and is transmitted efficiently from human to human in large-droplet aerosols. Both *B. anthracis* and variola virus encode a large number of virulence genes that are responsible for their capacity to infect and sometimes kill humans.

BOTULINUM NEUROTOXIN

The presence of a toxin gene may contribute to the pathogenicity of an organism. In some cases, the acquisition of a toxin genetic element may transform a relatively benign bacterium into a potent threat, as in the case of *Clostridium botulinum*. Although small-molecule toxins are found in shellfish or fungi and peptide toxins are found in venom from animal species, we will focus on the larger-protein toxins produced by bacteria and some plants. For the purpose of this appendix, we define a toxin as any protein that has a deleterious effect on health and fitness when ingested.

Seven immunologically distinct and extremely potent protein neurotoxins are produced by *C. botulinum*: A, B, C, D, E, F, and G. Types A, B, and E are most frequently associated with human disease, and types F and G are less often reported. Types C and D are associated with disease in fowl. The BoNTs are all expressed as single polypeptides that are posttranslationally proteolyzed to give a heavy chain and a light chain that are linked by a disulfide bond. The heavy chain is responsible fo r toxin adherence to the cell surface and translocation of the light chain into the cytosol. The light chain contains the zinc protease active site that is responsible for cleavage of the soluble *N*-ethylmaleimide-sensitive-factor attachment protein receptor (SNARE). BoNTs bind ganglioside receptors on the neuronal cell surface in the presynaptic terminals and, by their action as metalloendoproteases, selectively cleave proteins involved in the neuroexocytosis apparatus; this results in inhibition of acetylcholine release at the myoneural junctions and later flaccid paralysis that can lead to respiratory arrest.

The enzymatic active site of these toxins has been well characterized. A crystal structure of BoNT/A demonstrated that zinc is coordinated by the HisGluXXHis motif in the active site alpha helix through His-222 and His-226 and by Glu-261 (Lacy, Tepp et al. 1998). Site-directed mutagenesis experiments demonstrated the importance of these residues for toxin activity (Fujii, Kimura et al. 1992). Because the zinc binding site of this class of toxins is so well conserved (Fujii, Kimura et al. 1992) it would seem possible to predict from the gene sequence whether a bacterium is able to produce active toxin. However, data that argue against the use of BoNT sequence to predict activity comes from Agarwal et al. (Agarwal, Binz et al. 2005), who determined the structure of BoNT/E light chain with single amino acid substitutions outside of the canonical active site in regions adjacent to the catalytic residues. They found that the relatively minor change of glutamate to glutamine at position 335 rendered the enzyme unable to bind zinc. Other mutations in these regions also had large effects on the activity of the toxin. These findings serve to emphasize that biological toxin activity not only depends on the primary sequence but also on the three-dimensional structure of the folded protein.

BACILLUS ANTHRACIS

Disease

Bacillus anthracis is a gram-positive, spore-forming rod that is an etiological agent of pulmonary, cutaneous, and gastrointestinal anthrax. In cases of pulmonary anthrax, spores are inhaled into the airway and taken up by macrophages. The generally accepted model of pathogenesis is as follows: Germination of the spores begins to occur in alveolar macrophages and continues as the macrophages migrate to the mediastinal lymph nodes, where vegetative growth of the bacilli occurs. In the final stages of pulmonary anthrax, the bacilli disseminate throughout the body and cause death of the host. In cutaneous anthrax, spores penetrate the skin through a wound. The spores germinate into vegetative bacilli and proliferate locally to cause formation of an eschar. In some patients, the bacilli can migrate into and multiply in the bloodstream, spread to various organs, and, if left untreated, cause death of the host. Antibiotic treatment initiated before dissemination of the bacteria into the bloodstream during either type of infection dramatically reduces the onset of multiorgan failure and death. However, because the symptoms of early pulmonary anthrax mimic those of other infections, the disease is often not diagnosed until the patient is unable to recover, even with broad-spectrum intravenous antibiotic therapy. The bimodal life cycle of *B. anthracis* contributes extensively to the pathogenesis of disease. The organism survives outside the mammalian host as a spore that is resistant to heat, chemicals, desiccation, and so on. *B. anthracis* spores are coated with at least 34 chromosomally encoded proteins, many of which are immunogenic and protect the bacterium. The genes encoding the components of the spore are regulated by a complex temporal network of sporulation-specific sigma factors. Germination into vegetative bacilli occurs rapidly in the host in response to as yet unidentified signals.

Virulence Genes

Vegetative bacilli produce the two toxins edema factor (EF) and lethal factor (LF). The genes encoding the components of EF (*pagA* and *cya*) and LF (*pagA* and *lef*) are found on the 182-kb pXO1 plasmid. In addition, the bacilli are encapsulated by a poly-D-γ-glutamic acid capsule that is produced by enzymes encoded by genes (*capBCAD*) on the pXO2 virulence plasmid. Both plasmids are required for full virulence of *B. anthracis* strains. The genes encoding both the toxins and the capsule are activated by the master regulator of virulence AtxA, the gene for which is also on pXO1. The toxin genes are directly activated by AtxA, which also activates *acpA* and *acpB*, pXO2-encoded genes whose products activate the capsule operon. AtxA also exerts an effect on chromosomal genes that encode surface components, including the surface-layer proteins Sap and EA (*sap* and *eag*, respectively), by activation of the pXO1-encoded *pagR*; PagR represses *sap* and activates *eag*. Thus, at a minimum, a fully virulent strain of *B. anthracis* requires the genes necessary for spore, toxin, and capsule formation and a large array of regulatory factors and other genes that are chromosomally encoded. However, given the conservation of the *B. anthracis* genome over decades, if an isolate was identified as *B.*

anthracis by standard microbiologic means and if orthologous toxin and poly-D-glutamic acid capsular gene sequences were present, a presumption of virulence could be made. Proof of that assumption would require assessment of the relative lethality of spores prepared from the isolate in an animal model—mice, rabbits (better), or non-human primates (best).

VARIOLA VIRUS

Disease

There are three distinct clades of variola virus that coincide roughly with low, intermediate, and high case-fatality rate. Clade A was associated with an intermediate (8-12 percent) case-fatality rate. Clade B (variola minor) was associated with a low (below 1 percent) case-fatality rate and caused a disease referred to as amaas in Africa or alastrim in the Americas. Clade C (variola major) was associated with a high (16-30 percent) case-fatality rate and caused the disease classic or ordinary smallpox. Clinically, smallpox in an unvaccinated person has an incubation period of 7-19 days from the time of infection of the upper respiratory tract until the first symptoms of fever, malaise, headache, and backache occur. The characteristic rash then follows. The rash starts with papules, which sequentially transform into vesicles and then pustules; most of these lesions are on the head and limbs (often confluent) rather than on the trunk (centrifugal pattern). Lesions are 0.5-1 cm in diameter and can spread over the entire body. Once pustules have dried, scabs form and eventually desquamate during the next 2-3 weeks. The resulting feature of the cutaneous lesions is the formation of the classic pock scar that is apparent on the skin of surviving patients.

Virulence Genes

The protein coding regions of the variola virus genome is 96 percent identical at the nucleotide level to other orthopoxviruses. The majority of the sequence diversity occurs in the flanking regions of the genomes, which contain the virulence genes that target host apoptosis and the innate/immune response functions. Each poxvirus has 90 core genes and a unique complement of virulence genes, the latter of which determine in large part the unique biology of each orthopoxvirus species. These virus-specific features include the reservoir and incidental hosts, cell/tisariola virus encodes 71 virulence genes, of 41 of which something is known about function or location in the virion or infected cell; there is no experimental information on the remaining 30 genes. The putative roles of virulence genes in the natural life cycle of variola virus have been determined by the study of orthologous genes in other orthopoxvirus-animal models (such as vaccinia virus-mouse, ectromelia virus-mouse, and myxoma virus-rabbit). A number of the virulence genes encode cytokine-binding proteins, which can have varied specificity against ligands from different animal species. For example, the variola virus IL-18-binding protein has a higher affinity for mouse than for human IL-18; this suggests that adaptation to the human host does not always require or result in optimal specificity for the human ligand. Thus, the experimentally determined specificity of a gene product for host ligands may or may not support an inference of host biology. Some virulence genes target the same pathways. For example, NFκB is a key factor for transcription of host genes that mediate innate and immune responses and is targeted by a number of pathogens, including poxviruses (see Box I-1). Variola virus has at least five virulence genes thought to target that pathway: One gene product acts extracellularly by binding to IL-18, and the other four act intracellularly against signaling pathways or the NFκB complex. The importance of a gene for virulence is defined not only in the context of the host in which the virus replicates but also by the route of infection. For example, in the case of vaccinia virus infections of the mouse, 50 percent of 16 individual gene deletion or insertion mutants showed a phenotype distinct from that of controls in intranasal or intradermal infections but not both (Tscharke, Reading et al. 2002).

Even with the same complement of virulence genes, the case-fatality rate of variola virus isolates appears to vary, presumably because of subtle functional differences in individual genes or groups of genes, which are not understandable from sequence analysis. For example, amino acid differences in the coding-region sequences revealed that a consensus of 67 open reading frames (ORFs) distinguished clade A strains (middle-range case-fatality rates) from clade B strains (low case-fatality rates), and 15 ORFs distinguished the middle-range from the low case-fatality rate groups of clade C virus strains from Africa (Esposito, Sammons et al. 2006).

Variola virus is one of the few Select Agents that transmit efficiently from person to person during disease. That process does not occur at the primary site of infection; rather, it depends on a large number of virus-replication cycles in different cell types and tissues in the face of a rapidly activated, innate or adaptive immune response. High transmissibility depends on effective systemic virus spread and later establishment of sufficient foci of infection in the oropharyngeal mucosa to produce virus concentrations in the respiratory gases that are high enough to infect contacts. Although monkeypox virus differs from variola virus in only 12 virulence genes, it is unable to sustain transmission in human populations after introduction from an animal source. The genetic basis of the difference in transmissibility between monkeypox and variola virus is unknown, and there are no physiologically relevant animal models that could be used to answer the question.

In summary, many virulence genes are required for the full virulence of *B. anthracis* and variola virus, and a number of these genes are directly or indirectly contextual to the particular animal species, route of infection, cell or

BOX I-1
VariolaVirus Virulence Genes

	Predicted function/location		*Predicted function/location*
1	IMV-MP/Virulen. factor (Cop-A14.5)	**22**	Complement binding protein (Cop-C3)
2	Host-range (Cop-A51)	**23**	Guanylate kinase (Cop-A57)
3	Serpin 1,2,3 (3 paralogues)	**24**	TLR/IL-1 signaling inhibitor (Cop-A46)
4	Ubiquitin Ligase/Host defense (Bsh-75)	**25**	dUTPase (Cop-F2)
5	EGF Growth factor (Cop-11)	**26**	Ribonucl. Reduct. small subunit (Cop-F4)
6	NFκB inhibitor (Cop-N1)	**27**	Membrane prot. (Cop-F5)
7	Apoptosis inhibitor (Cop-F1)	**28**	Ribonucl. Reduct. large subunit (Cop-I4)
8	Superoxide dismutase-like (Cop-A45)	**29**	Thymidine kinase (Cop-J2)
9	IFN-gamma receptor (Cop-B8)	**30**	Fusion prot. (Cop-A27)
10	IFN-alpha/beta receptor (Cop-B19)	**31**	EEV prot. (Cop-A33)
11	P4c precursor (Cop-A26)	**32**	CD47-like prot. (Cop-A38)
12	IFN resistance/eIF2 alpha-like (Cop-K3)	**33**	Profilin homolog. (Cop-A42)
13	Chemokine binding protein (Cop-C23)	**34**	EEV prot. (Cop-A34)
14	EEV membrane prot. (Cop-E7)	**35**	Virulence (Cop-A41)
15	36kDa membrane prot. (Cop-F5)	**36**	IEV prot. (Cop-A36)
16	NFkB inhibitor (Cop-M2)	**37**	Membrane prot. (Cop-A43)
17	TNF receptor (CrmB; (Cop-C22)	**38**	Thymidylate kin. (Cop-A48)
18	NFkB inhibitor (Cop-K1)	**39**	DNA ligase (Cop-A50)
19	IL-18 BP (Bsh-D7)	**40**	Ser/Thr kin. (Cop-B1)
20	IL-1R antagonist (Cop-C10, C4)	**41**	EEV prot. (Cop-B5)
21	Put. Phostrans/anion transpt. (Cop-A49)		

( )—ORF designation of the prototype orthologue; abbreviations: Virulen.-virulence; Put. Phostrans/anion transpt—extracellular enveloped virus; prot.-protein; Put. Phostrans/anion transpt. prot-putative phosphotransferase/anion transport protein; ribonucl. Reduct.-ribonucleotide reductase; kin.-kinase; ser/thr-seerine/threonine; IEV-intracellular enveloped virus.

biochemical pathways. For poxviruses, the presence or absence of a virulence gene is not informative as to the infectivity of the virus in any animal species, including humans. Furthermore, in some situations, it is likely that subtle changes in the activity of one or more variola virus genes alone can affect pathogenicity. Those hypothesized subtle differences in activity cannot now be predicted from sequence analysis.

In the following section, a number of general mechanisms for the evolution of pathogenicity and sustainability in a host are described with an emphasis on the important observation that closely related pathogen species or strains can evolve to be pathogenic in a host in unique and sometimes unpredictable ways.

FILOVIRUS PATHOGENESIS

Disease

The filovirus family, Filoviridae, consists of two genera: *Marburgvirus*, which comprises various strains of the 1967 Lake Victoria marburgvirus (MARV), and the antigenically distinct *Ebolavirus*. Ebolaviruses were first discovered in 1976. The genus contains five species: Sudan ebolavirus (SEBOV), Zaire ebolavirus (ZEBOV), Ivory Coast ebolavirus (CIEBOV)), Bundibugyo ebolavirus, (BEBOV), and Reston ebolavirus (REBOV). Filoviruses are among the deadliest of all human pathogens, causing hemorrhagic fever with mortality that can approach 90 percent. Several reports indicate high seroprevalance in many areas of Africa. Assuming that the serologic tests were specific, either EBOV is endemic or there are a set of uncharacterized, cocirculating, nonpathogenic, antigenically cross-reactive viruses. In support of that idea, REBOV is not pathogenic in humans.

Filoviruses are filamentous, nonsegmented negative-strand RNA viruses that have about a 19-kb RNA genome, have a highly conserved gene order, and are surrounded by a helical nucleocapsid structure and a lipid bilayer that contains several virus glycoprotein spikes. These viruses probably target bats as reservoir species, although Marburg Reston is maintained in swine in the Philippines. In uninterrupted human-to-human transmission, nucleotide sequence changes are rare, except in an Angola outbreak characterized by targeted evolution in VP40 and VP24. It is possible that these viruses require minimal evolution for human replication and pathogenesis.

Early in infection, filoviruses target cells of mononuclear lineage, notably macrophages, monocytes, and dendritic cells (DCs), but not lymphocytes. Infected monocytes release inflammatory cytokines, such as TNF-α, whereas DCs are anergic (characterized by limited cytokine production, DC maturation, and diminished antigen presentation for T-cell activation). Neutrophils rapidly become activated, most likely by viral glycoprotein interaction with the TREM-1 ligand, and this results in increased cytokine production. Massive bystander apoptosis of natural killer cells and lymphocytes occurs intravascularly and in lymphoid organs. Innate immunity and adaptive immunity are delayed during filovirus infections, and this allows increased virus replication and disease exacerbation as many cells (such as monocytes and neutrophils) are continuously triggered to release cytokines. As infection increases, filoviruses infect a wide variety of cells and organs, including the liver and endothelial cells. Dysfunctions in hemostasis occur as a consequence of hepatic damage and the release of TNF-α and other proinflammatory cytokines. In fatal cases, death occurs 6-16 days after the onset of symptoms, usually because of multiorgan failure and coagulopathy that results in disseminated intravascular coagulation and shock. Hemorrhagic disease occurs in about 25-45 percent of the patients and is most likely triggered by immune-mediated mechanisms.

Virulence Determinants

Filovirus virulence determinants include viral proteins that antagonize adaptive and innate immune responses, suggesting a role for inflammation in resistance and disease. Extensive replication of filoviruses in primates is regulated by two key viral proteins that antagonize host interferon responses. VP35 inhibits the activation of transcription factor IRF3 by binding to dsRNA and inhibiting retinoic acid induced gene-I (RIG-I) signaling. VP35 also interferes with the activation of the dsRNA-binding kinase, PKR. Mechanistically, VP35 expression likely augments the conjugation of a small ubiquitin-like modifier (SUMO) protein to IRF3/IRF7 through TLR and RIG-1 signaling, leading to increased inhibition of IFN transcription by IRF3/7. Thus, VP35 activates a normal negative feedback loop that regulates IFN signaling to weaken host innate immunity. In contrast, VP24 inhibits the cellular response to exogenous IFN by interacting with karyopherin α1, preventing the nuclear accumulation of tyrosine phosphorylated STAT1 and STAT2.

The GP glycoprotein, including its soluble sGP form, also functions as a major virulence determinant, playing an important role in virus attachment and entry, cell rounding, cytotoxicity and down-regulation of host proteins. GP toxicity is thought to be mediated by a dynamin-dependent protein trafficking pathway and a ERK mitogen activated protein kinase pathway. Importantly, exposure of primate PBMCs to select ZEBOV, or MARV GP peptides or inactivated ZEBOV resulted in decreased expression of activation markers on CD4 and CD8 cells; CD4 and CD8 cell apoptosis; blocked CD4 and CD8 cell cycle progression; decreased interleukin (IL)-2, IL12-p40 and IFN-gamma; but increased IL-10 expression. Thus, GP likely encodes an immunosuppressive motif that likely antagonizes adaptive immunity during infection.

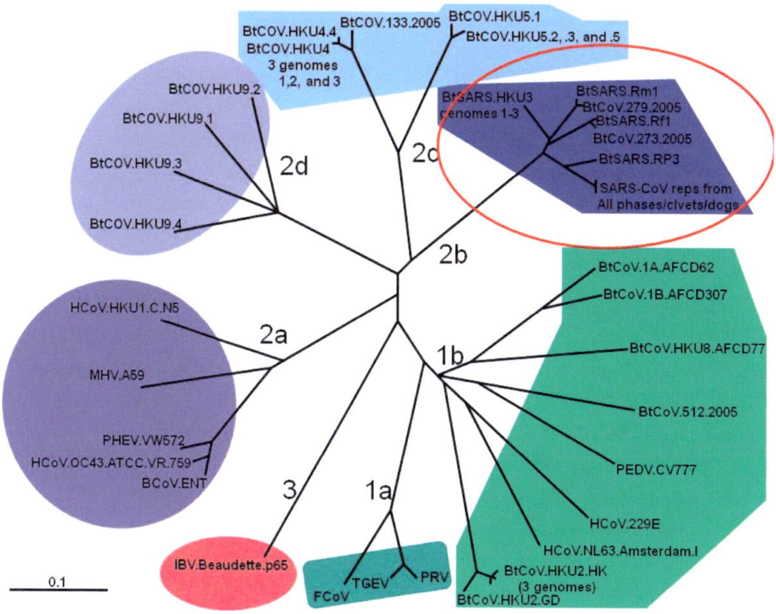

FIGURE I.1 Coronavirus phylogeny.

SARS CORONAVIRUS

Pathogenesis

Coronaviruses encode a ~30 kb single-stranded positive polarity RNA genome that is wrapped in a helical nucleocapsid composed of multiple copies of a nucleocapsid protein and surrounded by a lipid envelope bearing three or more glycoprotein spikes. The virus family is divided into group 1 (alpha coronaviruses), 2 (beta coronaviruses), and 3 (gamma) coronaviruses based on sequence homology (See above Fig—Coronavirus Phylogeny) Coronavirus phylogeny and biology are characterized by frequent host-shifting events, including animal-to-human (zoonosis), human-to-animal (reverse zoonosis) or animal-to-animal. Over the past 30 years, several coronavirus cross-species transmission events as well as changes in virus tropism have given rise to new significant animal and human diseases. Most notably, severe acute respiratory syndrome (SARS), a human lower respiratory disease that was first reported in late 2002 in Guangdong Province, China, quickly spread worldwide over a period of four months. The virus infected over 8,000 individuals, killing nearly 800 before it was successfully contained by aggressive public health intervention strategies. The etiological agent of SARS (SARS-CoV) was determined to have crossed into human hosts from zoonotic reservoirs including bats as well as Himalayan palm civets (*Paguma larvata*) and raccoon dogs (*Nyctereutes procyonoides*) that were sold in exotic animal markets in China. Of note, another human coronavirus (HCoV-229E) likely emerged from African bat coronavirus lineages some 200 years earlier while HCoV OC43 likely emerged from closely related bovine coronaviruses about 100 years ago. SARS-CoV was recently proposed as a new select agent based primarily on its high virulence and transmission potential in human populations and the lack of effective vaccines and therapeutics. This recommendation models many of the difficulties in using sequence-based criteria for determination of virulence potentials. SARS-CoV is a group 2b coronavirus, which includes closely related civet and raccoon dog strains (>99 percent sequence identity) as well as more variant bat coronaviruses, HKU3 and RB3. Protein sequence identity is greater than 95 percent across most of the genome of human epidemic SARS-CoV and bat group2b coronaviruses although the S glycoproteins are only 80-90 percent identical. It has been proposed that bat coronaviruses were the progenitor strains for all group 1 and group 2 coronaviruses and genome homologies range from 43 to >90 percent amino acid identity. Obviously, some viral genes are more highly conserved than others.

Age was a major virulence determinant as mortality rates were less than 1 percent for individuals below 21 years of age, but >50 percent in individuals greater than 65 years of age. The predominant pathological features of SARS-CoV infection in the human lung included diffuse alveolar damage (DAD), hyaline membranes, atypical pneumonia with dry cough, persistent fever, progressive dyspnea and sometimes abrupt deterioration of lung function. Virus infection primarily targeted ciliated epithelial cells and type II pneumocytes in the lung as well as epithelial cells in the intestine. Major pathologic lesions include inflammatory exudation in the alveoli and interstitial tissue with hyperplasia of fibrous tissue and fibrosis. Two phases of disease were identified during SARS-CoV infection in humans. Acute respiratory distress syndrome (ARDS) develops within the first 10 days with DAD, edema, and hyaline membrane formation. After the acute phase, an organizing phase DAD with increased fibrosis is observed. Increasing age, male sex, presence of comorbid conditions, high early viral RNA burdens, and high lactate dehydrogenase levels are associated with greater risk of death. In serum, dynamic changes in cytokine levels have been reported following SARS-CoV infection including increases in IFN-γ, IL-18, TGF-β, IL-6, MP-10, MCP-1, MIG and IL-8, but not TNF-α, IL-2, IL-4, IL-10, IL-13 or TNFRI. The data suggested that an IFN-γ-related cytokine storm might be involved in the immunopathological damage noted in SARS patients.

		ORF 1A												1B		Spike																	M	Orf 3a				Orf 8
	AA change	A-S	V-A	I-T	P-L	L-I	L-F	C-Y	L-S	C-W	V-A	V-A	V-A	D-E	R-K	G-D	N-K	S-L	I-T	T-K	R-K	F-S	N-K	T-S	S-P	L-S	S-L	T-A	V-A	Y-D	T-A	E-K	G-S	F-I	C-S	H-Y	C-G	
Phase	AA position	549	1021	1121	1136	1663	2116	2222	2269	2746	2971	3047	3072	1389	2532	77	227	239	244	261	344	360	479	487	607	665	701	743	754	778	894	1163	5	7	81	93	121	Δ or +
Animal	SZ16	S	A	T	L	I	F	Y	S	W	A	A	A	E	K	D	K	L	T	K	R	S	K	S	P	S	L	A	A	D	A	E	S	I	S	Y	G	+29
	HC/SZ/61/03	A	V	T	P	I	L	C	L	C	A	A	A	E	R	D	K	S	T	T	R	S	R	S	S	S	S	T	V	D	T	E	S	I	S	H	G	+29
Early	GZ02	A	V	T	P	I	F	Y	L	W	A	A	A	E	K	D	N	L	T	T	R	F	N	T	S	L	S	T	V	D	T	E	G	F	C	H	C	+29
Middle	CUHK-W1	A	V	I	P	L	L	C	L	C	V	A	A	E	R	D	N	S	T	T	K	F	N	T	S	L	S	T	V	Y	T	K	G	F	C	H	C	Δ29
Late	Urbani	A	V	I	P	L	L	C	L	C	V	V	V	D	R	G	N	S	I	T	K	F	N	T	S	L	S	T	V	Y	T	K	G	F	C	H	C	Δ29

FIGURE I.2 Conserved sequence alterations associated with expanding phases of the SARS-CoV epidemic. Hypothetically, the pandemic strains of SARS evolved from the virus isolated from civet cats (SZ16, HC/SZ61/03). GD03 is a very early human isolate (Dec 22, 03) very similar to HC/SZ61/03. The GZ02 sequence is representative of early strains (e.g., HGZ8L1-A, etc.) detected in a patient in Guangzhou (the capital of Guangdong). CUHK-W1 and related viruses (Guangdong), are reasonable precursors to viral strains that evolved from early Guangdong isolates and is representative of the middle phase of the epidemic. Urbani is representative of late phase isolates that occurred after the Metropol Hotel superspreader event. Residues in green are indicative of the civet cat alleles, pink are mutations that occurred early during civet cat-human transmission, dark blue during the middle phase and gray during the late phase. We will build viruses encoding these mutational profiles.

Virulence Determinants

SARS-CoV Cross-Species Transmission. The SARS-CoV outbreak is unique in that a chronological set of sequence changes were obtained, providing precise sequence signatures associated with expanding waves of the global epidemic. By comparing the earliest full length human isolate (GZ02; early December 2002) to civet cat isolates SZ16 and HZ/SZ/63 and key strains epidemiologically linked to an expanding epidemic, 12 amino acid changes in ORF1a, 2 in ORF1b, 17 in the S glycoprotein, 4 in ORF3a, 1 in the M glycoprotein and the ORF8 29 nt deletion were identified that may have allowed for increased replication, transmission, and pathogenesis in human hosts (See Figure above). In general, civet/raccoon dog-related strains and strains identified in sporadic human cases prior to the onset of the epidemic were thought to be significantly less transmissible and pathogenic than those identified during the early, middle and late stages of the epidemic. Except for a couple of mutations in S and M which either promoted entry into human cells or promoted efficient egress, respectively, the role of most mutations in the expanding epidemic still remain unclear. Importantly, civet and raccoon dog strains do not replicate in human cells, despite having greater than 99 percent sequence identity. Only two or three mutations are needed for to promote efficient replication in human airway cells, yet animal strains whose S RBD recognize hACE2 receptors are only weakly pathogenic and require extensive adaptation in S and elsewhere in the genome. Current proposals to include "SARS-CoV" as a potential select agent are unclear regarding the disposition of the closely related but human host range-restricted, civet and raccoon dog SARS-like strains, sporadic human strains identified in 2004 that are mostly zoonotic in origin, as well as bat SARS-CoV-like strains in this classification scheme. More importantly, given the close homology but receptor mediated restriction and/or limitation in hACE2 recognition by these animal-origin strains, defining SA status by global genome sequence homology seems arbitrary and not grounded in any rational scientific method.

At ~180 kDa in mass, the S glycoprotein is a trimer in the virion and organized into two subunit domains, an amino-terminal S1, which contains the ~200-aa receptor binding domain (RBD), and a carboxy-terminal S2, which contains the putative fusion peptide, two heptad repeat (HR) domains, and a transmembrane domain (TM). This domain organization groups the CoV S spike glycoprotein with other class I viral fusion proteins, such as Influenza HA, HIV-1 env, SV5 F, and Ebola Gp2. The RBD of Spike is generally acknowledged as the principal determinant of coronavirus host range and two key mutations in S, K479N and S487T, were shown to be responsible for promoting efficient interaction with either the civet or human ACE2 receptor, respectively. Given these findings, it seems that reasonable criteria could be developed to use sequence-based criteria to categorize "human and animal" strains for oversight purposes. However, multiple pathways of S-driven host range expansion exist, complicating sequence based predictions of host range expansion. For example, re-adaptation of the civet SZ16 strain to human airway epithelial cells identified a different set of key "humanizing" mutations at K479N, Y442F and L472F. In addition, K479T and K479I also allow for efficient S RBD interaction with hACE2 demonstrating that multiple genetic mutation pathways exist, which would allow for differential recognition of civet and/or human ACE2 receptors. Importantly, the SARS-CoV RBD is highly plastic and can be recombined into closely related group2b bat and animal strains, promoting efficient growth in human airway epithelial cells. These strains are significantly less efficient at replicating in mouse models of human disease, suggesting that additional adaptation and mutation would be required to promote efficient disease potential in humans. In fact, the S ectodomains are readily interchangeable between group 1 and 2 coronaviruses and these recombinant viruses grow efficiently. Given the high recombination frequency noted in mixed coronavirus infection, recombination driven "host shifts" are likely common in coronavirus phylogenies. Finally, by completely novel mechanisms, as little as 2-4 mutations in some coronavirus fusion peptide and HR1 domains have promoted host range expansion into multiple species in cell culture. Although the mechanism is unclear, it represents an entirely different pathway to host range expansion. Clearly, the extensive plasticity and existence of multiple interaction networks by which S can augment receptor(s) interactions and entry, makes sequence based predictions of host range specificity difficult if not usually impossible.

Mutations that effect virus replication efficiency also must be considered an important virulence determinant. It is likely that a number of mutations that occurred during the SARS-CoV outbreak selected for more efficient virus-human host interactions that promoted efficient replication, gene expression and release. Consonant with this hypothesis, mutations in virtually any essential gene attenuates virus replication and pathogenesis in vivo, demonstrating the importance of efficient virus growth in disease progression. In general, this seems to be a common truism for most RNA virus genomes. Nevertheless, SARS-CoV also encodes a number of proteins that strongly antagonize the innate immune sensing/signaling pathways and which likely function as important virulence determinants that regulate pathogenesis. Among the 16 SARS-CoV ORF1 replicase proteins, nsp1, nsp3 (PLP-papain like protease), nsp7 and nsp15 all encode strong interferon antagonist activities that either block type 1 interferon and/or NFκB sensing/signaling pathways. These four non-structural proteins are encoded in all coronavirus genomes but are highly variable in primary amino acid sequence making it difficult to predict whether similar innate immune antagonism activities are universally encoded in all Coronaviridae genomes. In addition, nsp1 is reported to block host translation and promote host, but not viral, mRNA degradation, and nsp1 and PLP mutations have been described which attenuate innate immune antagonism activities and/or attenuate virus pathogenesis. However, even in closely related coronaviruses, these homologous proteins may or may not encode similar innate immune antagonism functions, making it difficult to predict activity based on sequence homology and biologic precedent. For example, the HCoV NL63 and SARS-CoV nsp3 encoded PLP domains are highly variable, differing by over 80 percent in amino acid identity, yet both likely form similar structures and antagonize IFN signaling. It is not clear whether common sequence motifs regulate this activity. In addition to the non-structural proteins encoded in ORF1, antagonist activities are also encoded in ORF3b, ORF6 and the N protein, which either block type 1 interferon sensing and/or signaling by targeting novel cellular components essential for signal transduction. Except for ORF6 from the closely related Bt-SARS-CoV HKU3, it is not clear whether closely related group2b homologs also encode similar IFN antagonist activities. Several SARS-CoV proteins (e.g., S, M, ORF3a, ORF6, ORF7, N, etc.) are pro-apoptotic and likely contribute to virus cell killing and pathogenesis as well.

The nsp1 proteins of coronaviruses are highly variable, both in terms of size and amino acid sequence variation. The SARS-CoV NSP1 is a 20kd protein that is localized to the cytoplasm of infected cells. In some reports, the NSP1 was able to block IFN-β mRNA induction but did not antagonize the IRF3 signaling pathway. NSP1 expression degraded not only IFN-β mRNA but also several endogenous cellular mRNAs as well and inhibited in translation of cellular mRNAs. As SARS-CoV infected cells also degraded cellular mRNA, these authors proposed that nsp1 degradation of host mRNA is an important mechanism of blocking host antiviral defenses. However, other work suggests that SARS-CoV NSP1 inhibits the signal transduction pathways involving IRF3, STAT1 and NF-kB. Interestingly, these authors did not observe the mRNA degradation phenotype seen in the previous studies.

ORF6 is a 63aa ER/Golgi membrane protein that has its C terminal tail facing the cytoplasm and its N terminus either in the ER lumen or associated with the ER membrane. ORF6 blocks the nuclear import of the STAT1/STAT2/IRF9 and STAT1/STAT1 complexes in the presence of IFNβ or IFNγ treatment, respectively, and resulted in the reduction of STAT1-dependent gene induction. The C terminal 10aa were critical for this import block; mediated by a recruitment of nuclear import factors to the ER/Golgi membrane. Karyopherin alpha 2 (KPNA2) specifically binds to the C-terminal tail of ORF6, retaining KPNA2 at the ER/Golgi membrane. This interaction subsequently recruits Karyopherin beta 1 (KPNB1) to the ER/Golgi membrane as well. The recruitment of KPNB1 onto membrane complexes limited the bioavailability of KPNB1, an essential component for the nuclear import of STAT1 complexes as well as other cargo. Consequently, ORF6 blocked nuclear import of the STAT1 complexes. ORF6 may affect other signaling pathways since KPNB1 is a common component of the classical nuclear import pathways. In recombinant viruses, ORF6 has also been shown to increase pathogenesis of the normally non-lethal MHV-A59 virus. In vitro studies with the MHV/ORF6 virus show that ORF6 expression increased the virus production from cells compared to WT virus. Recently, Hussain et al showed that ORF6 expression blocked proteins containing classical import signals but not proteins that use non-importin nuclear import mechanisms (Hussain, Perlman et al. 2008). MHV's accessory proteins have not been implicated in impacting the innate immune response however they may clearly play a role in MHV's capacity to evade the innate immune sensing proteins as deletion of some, but not all impact in vivo pathogenesis. The clear association of SARS-CoV ORF6 and Ebola VP24 with the host nuclear import pathway identifies nuclear import as a key site by which highly pathogenic viruses regulate the intracellular environment. By modulating the kinetics of nuclear import during infection, the virus controls innate immune, adaptive immune, apoptotic and cell stress signaling networks. While the mechanism may vary, it would be surprising if other pathogenic viruses don't modulate the same pathways as seen in SARS-CoV during their infection process.

The SARS-CoV and MHV nucleocapsid protein have also been shown to affect different aspects of the innate immune response and appear to modulate several signaling pathways in the cell. Many of these studies were performed using over expression constructs in isolation and have not been confirmed in the context of virus infection. He et al. showed that N protein is able to induce AP-1 signaling in vitro. SARS-CoV N was able to block the induction of reporter gene expression from an IFN-β promoter and also block NF-kB signaling. These data indicate that the N protein is also able to inhibit an ISRE promoter in response to Sendai infection but not IFN-β treatment. The mechanism by which this inhibition is occurring is unknown and under investigation. The MHV N protein has been shown to inhibit the activation of PKR, a strongly antiviral protein, in the cytoplasm from poxvirus vectors and in MHV-infected cells. PKR activation normally leads to a block in protein synthesis by phosphorylating the alpha subunit of the translation factor eIF2. While N does not itself prevent PKR activation, it alters PKR's function such that it no longer signals properly. The N proteins between these two distinct group 2 coronaviruses are quite conserved, so it will be interesting to determine if they both encode overlapping functions during infection or whether they mediate distinct inhibitory mechanisms.

Appendix J

Pathogenicity Acquisition

The following sections discuss mechanisms, and provide examples, of how a microorganism may acquire pathogenicity.

GENE GAIN

Generation of Multidrug Resistant Bacterial Pathogens

The emergence of multidrug resistant bacterial pathogens, including *Staphylococcus aureus*, *Mycobacterium tuberculosis*, and *Pseudomonas aeruginosa*, highlights the importance of understanding the mechanisms by which antibiotic resistance is transferred among different genera of bacteria. The acquisition of resistance to multiple antimicrobial agents is a hallmark of methicillin-resistant *S. aureus* (MRSA), organisms that can cause life-threatening skin and soft tissue infections, endocarditis, and pneumonia. The SCC*mec* cassette, which confers resistance to methicillin and other β-lactams, is chromosomally-encoded but is transferred by a mobile genetic element. Mobile genetic elements, including transposons, plasmids, and bacteriophage, are the most common sourccs of acquired antibiotic resistance in MRSA. In addition, spontaneous mutations in chromosomally encoded genes can confer resistance to antibiotics (e.g., *rpsL* and streptomycin).

The two types of MRSA circulating within the human population are community-acquired (CA-MRSA) and healthcare-associated (HA-MRSA) MRSA. HA-MRSA strains have been propagated in healthcare facilities in the face of antibiotic pressure and continue to survive and multiply as a result of acquisition of multiple elements conferring resistance to antibiotics. In contrast, CA-MRSA strains have been propagated in the absence of extreme selective pressure and have acquired resistance to fewer antibiotics. Both types of MRSA maintain chromosomal mutations that confer resistance to some therapies; these mutations are selected for and maintained in healthcare environments. HA-MRSA strains cause disease in immunocompromised individuals and are rarely isolated from patients with MRSA infections that are not linked to healthcare settings. In contrast, CA-MRSA strains typically cause infection in immunocompetent and otherwise healthy people. The mechanisms underlying this difference are still unclear and are likely multifactorial. Some evidence suggests that the acquisition of multiple antibiotic-resistance determinants by HA-MRSA strains actually reduces the "fitness" of the organism, which makes them unable to colonize and/or infect otherwise healthy people.

Toxins

Protein toxins in bacteria are often associated with mobile genetic elements such as phages and plasmid DNA. For this reason many toxin genes have spread between species by horizontal gene transfer. Production of active toxin sometimes requires the presence of accessory proteins for posttranslational modification and/or export. Thus the detection of a toxin gene is not necessarily predictive of virulence. These toxins may be grouped into two broad categories, structural and enzymatic. Structural toxins produce an effect solely through interactions with the target cell while enzymatic toxins catalyze a specific reaction that has an effect on the cell.

Acquisition of Shiga toxin-encoding bacteriophage. The emergence of Shiga toxin-producing *E. coli* strains in the early 1980s provides a clear example of how common bacteria may acquire bacteriophages that encode virulence traits such as bacterial toxin genes. This horizontal transfer of genetic material among bacteria via phage infection can result in the rapid evolution of new pathogens. The *E. coli* serotype O157 cited above was previously unknown as a pathogen in humans; however, the incorporation of genes homologous to those that encode Shiga toxin in the related agent of bacillary dysentery, *Shigella dysenteriae* type 1, generated a strain that produces hemorrhagic colitis and the life-threatening hemolytic uremic syndrome in humans.

Shiga toxins are AB_5 toxins (one polypeptide chain that has enzymatic activity and 5 cell-binding subunits) that are among the most potent toxins known. They kill sensitive cells by shutting down protein synthesis in a manner identical to that of ricin. Indeed, Shiga toxins, like ricin, are classified as SAs. In addition to *E. coli* O157, several other serotypes of *E. coli* and some related species including *Shigella sonnei*, *Aeromonas hydrophila*, and *Enterobacter cloacae* have been described that encode related Shiga-type toxins within lysogenic (chromosomally-integrated) bacteriophages and cause disease in humans. Bacteriophages are thought to be the most plentiful infectious forms on Earth.

The ubiquitous nature of these agents, and the frequency with which they recombine with one another during bacterial infection, provides an enormous opportunity for the exchange of phage and bacterial DNA. Furthermore, during the lytic cycle of bacteriophage replication, numerous phage particles are generated and released upon lysis of an infected bacterium. This provides the opportunity for additional bacterial infections and transfer of phage-encoded virulence factors. Indeed, transfer of Shiga toxin-bearing phages from pathogenic strains of *E. coli* to "bystander" *E. coli* strains that constitute normal intestinal flora has been observed during experimental infections in animals and certainly occurs in nature.

Cholera toxin acquisition via toxin-coregulated pilus mediated CTXΦ uptake. *Vibrio cholerae* is the etiological agent of cholera, a disease that manifests as either mild self-limited diarrhea or potentially fatal watery diarrhea and vomiting. The hallmark watery diarrhea results from intoxication with cholera toxin (CT), an AB_5 toxin that binds to the ganglioside GM_1 on the surface of intestinal epithelial cells. Five B subunits form a pentameric pore to facilitate entry of the A subunit into the cell; the internalized A subunit transfers ADP-ribose to a G protein that activates adenylate cyclase, which leads to increased cAMP production and hypersecretion of water and electrolytes. The genes encoding CT (*ctxAB*) are encoded on CTXΦ, a filamentous phage that is thought to be transferred from one bacterium to another during infection of the small intestine (McLeod, Kimsey et al. 2005). The *V. cholerae* surface receptor for CTXΦ is the toxin-coregulated pilus (TCP), a type IV bundle-forming pilus whose synthesis is encoded by genes found on a PAI on the *V. cholerae* large chromosome. In addition to its function as the receptor for CT, TCP is factor that is required for *V. cholerae* intestinal colonization. Once CTXΦ is internalized into the bacterial cell, it integrates into the host genome using host machinery. Transcription of phage genes required for phage replication and virion production is repressed during lysogeny. Expression of *ctxAB* and the *tcp* loci is controlled by a complex network of regulatory elements known as the ToxR regulon (Matson, Withey et al. 2007). Unlike most phage that encode virulence factors, CTXΦ is never excised from the chromosome; the mechanism underlying this phenomenon is not known. Instead, phage replication occurs during genome replication, and new virions are produced as a result of transcription and translation of integrated chromosomal phage genes when transcription is derepressed under stress conditions (McLeod, Kimsey et al. 2005). Virions are secreted using machinery encoded by both the phage and the host; once secreted into the extracellular milieu, phage can bind to the TCP of a susceptible host cell and be internalized. Identification of the *ctxAB* genes in the *Vibrio cholerae* genome would suggest that the particular strain could cause water diarrhea; however, regulation of *ctxA* is sufficiently complicated that mere identification of the toxin genes would not be predictive of virulence.

Capsule Switching in *Streptococcus Pneumoniae*

Streptococcus pneumoniae is an etiologic agent of otitis media, pneumonia, meningitis, and septicemia. The incidence of invasive pneumococcal disease has decreased since the introduction of two vaccines that protect against infection with *S. pneumoniae*. The vaccines are composed of capsular polysaccharides from 7 (Prevnar™) or 23 (Pneumovax®) of the most commonly isolated serotypes of *S. pneumoniae*. To date, 91 serotypes of *S. pneumoniae* have been described based upon their unique capsule structures (Park, Pritchard et al. 2007). Recombination of capsule loci, or capsular switching, is a common mechanism utilized by *S. pneumoniae* to evade host defenses. The regions flanking the capsule loci are very similar among capsular serotypes, and this sequence characteristic facilitates homologous recombination of capsule genes acquired through horizontal gene transfer in this naturally competent organism. Routine vaccination of young children with Prevnar™ has led to a significant decrease in the incidence of pneumococcal disease caused by the 7 serotypes in the vaccine. However, because *S. pneumoniae* is readily able to undergo horizontal transfer of its capsule genes, non-vaccine serotypes of the pneumococcus have emerged as important causative agents of invasive disease. Thus, the bacteria are rapidly adapting to an alteration in host susceptibility.

O-Antigen Evolution in *Salmonella*

Salmonella species are masters of adaptation to the host environment (Kingsley and Bäumler 2000). Upon infection with a *Salmonella* species to which it is susceptible, the host will develop symptoms of salmonellosis, which, in humans, is often characterized by gastroenteritis. The majority of humans will resolve the symptoms of *Salmonella typhimurium* subsp *typhi* infection within a day or two. In a small subset of cases, the bacteria enter the bloodstream to cause bacteremia. Ultimately, *Salmonella* migrates to the gall bladder, which is an immunologically protected environment, and resides there for years in a chronic colonization state. Occasionally *Salmonella* are released back into the intestine when bile is secreted, which allows for shedding of the organism and transmission to a new host.

The vast majority of mammals that become infected with *Salmonella* spp. produce antibodies against the lipopolysaccharide (LPS) O-antigen; O-antigen is the highly variable terminal oligosaccharide structure that forms the basis for the serogroups of *Salmonella*. Once a host develops antibodies against the O-antigen, the host is protected from reinfection with the same or a different strain that harbors the same O-antigen. Within a population, infection followed by protective antibody generation eventually reduces the virulence of strains that harbor a particular O-antigen to the point that the organism must evolve in order to survive. The genes that encode the enzymes required for O-antigen synthesis routinely undergo horizontal gene transfer in the face of host adaptation. When the O-antigen structure changes, the population becomes susceptible to infection with the strain that expresses the newly acquired O-antigen, and the cycle begins anew.

GENE LOSS

Shigella spp

The evolution of a non-pathogenic bacterial species into a genetically related but pathogenic species typically occurs as a result of the acquisition of genes encoding virulence factors, often via transfer of pathogenicity islands. However, loss of a gene or genes can also be critical to the capacity of a non-pathogenic organism to be converted to one that causes serious disease. Such is the case for the evolution of non-pathogenic *Escherichia coli* into pathogenic *Shigella* spp. and the genetically and clinically similar enteroinvasive *E. coli* (EIEC). Laboratory strains of *E. coli* contain a gene, *cadA*, that encodes lysine decarboxylase, an enzyme that catalyzes the decarboxylation of lysine to form cadaverine. The *cadA* locus has been deleted from *E. coli* during the evolution of the pathogenic *Shigella* spp. and EIEC. While cadaverine does not affect the invasive capacity of *Shigella* spp. and EIEC, it completely inhibits enterotoxin activity of *Shigella* spp.(Maurelli, Fernandez et al. 1998). Thus, the loss of *cadA* enhanced the virulence of the *Shigella* spp. and EIEC.

Y. pestis

Gene loss was a key mechanism in the evolution of *Yersinia pseudotuberculosis* to *Y. pestis*. *Y. pseudotuberculosis* is a free-living bacterium that causes mild gastroenteritis in humans and animals. In contrast, *Y. pestis* is an extremely virulent organism that causes bubonic and pneumonic plague. The host range for the two organisms differs considerably in that *Y. pestis* requires a vector (the flea) for transmission between mammalian hosts, while *Y. pseudotuberculosis* is transmitted freely between mammalian hosts. Genomic analyses of representative strains from each species demonstrated that, despite the difference in host range, the two organisms are very closely related at the nucleotide level, and that *Y. pseudotuberculosis* is the most recent ancestor of *Y. pestis* (Chain P). However, and perhaps owing to the need to adapt to different host niches, up to 13 percent of the *Y. pestis* chromosome is inactivated.

GENE MUTATION

Filoviruses. The Marburg and Ebola filoviruses and the SARS-CoV coronavirus replicate poorly in adult mice. Mouse adaptation of these viruses has

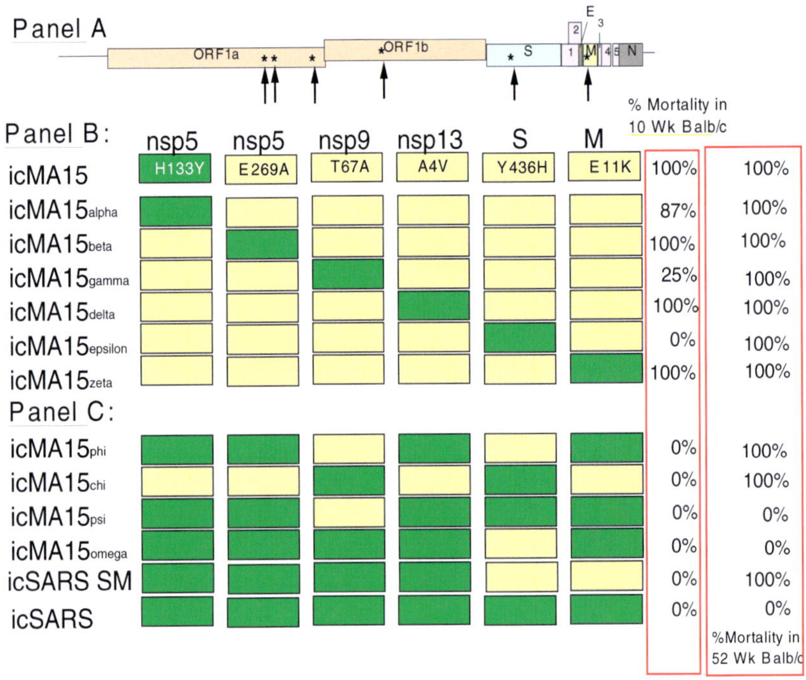

FIGURE J.1 Genome organization of icMA15 (**Panel A**). A series of chimeric icMA15 viruses were generated with different subsets of 5 icMA15 mutations in the icSARS-CoV WT genome, noting that wildtype positions in nsp9 and the S gene attenuated virulence in young but not aged animals (**Panel B**). Smaller combinations of the MA15 mutation set showed that 2-set nsp9/S or S/M were sufficient to kill aged animals only (**Panel C**).

provided an opportunity to identify mutation sets that are responsible for high virulence and cross species adaptation, although the approach is often limited by the isogenic host background as compared with outbred populations in natural settings. By blind serial passage in progressively older mice, mouse-adapted Ebola (MA-ZEBOV) acquired eight amino acid changes that were associated with high virulence, as well as a variety of silent changes. Using recombinant viruses, introduction of wild-type alleles into MA-ZEBOV VP24 and/or the NP protein were significantly associated with decreased virulence in the BALB/c mouse model. Moreover, efficient MA-ZEBOV replication in mice required the mouse adapted NP and VP24 genes, the latter is a potent inhibitor of type I interferon signaling. Recombinant viruses encoding the MA-VP24 mutation grew extremely efficiently in IFN-treated mouse peritoneal macrophage cell lines. The data indicate that VP24 type I interferon antagonist activity is a significant virulence factor for ZEBOV replication and pathogenesis in mammals, although the molecular mechanisms governing its *in vivo* function remain unclear (Ebihara, Takada et al. 2006). Ebola VP24-mediated inhibition of cellular responses to IFNs correlates with the impaired nuclear accumulation of tyrosine-phosphorylated STAT1; mostly likely by VP24 interactions with karyopherin alpha1, but not with karyopherin alpha2, alpha3, or alpha4. It is possible that the mouse adapted mutations in VP24 confer increased targeting and antagonism of murine Karyopherin alpha 1 protein import machinery, although direct proof is lacking. Serial passage of Marburg virus was first done in immunodeficient (SCID) mice, and then in immunocompetent mice. *In vivo* adaptation has also resulted in a mouse-adapted strain MARV-Ravn. MARV-Ravn also caused uncontrolled viremia and high viral titers in the liver, spleen, lymph node, and other organs; profound lymphopenia; destruction of lymphocytes within the spleen and lymph nodes; and marked liver damage and thrombocytopenia in BALB/c mice. Sequence analysis of the mouse-adapted MARV-Ravn strain revealed differences in 16 predicted amino acids from the progenitor virus, although the exact changes required for adaptation are unclear at this time.

Serial passage of the late phase epidemic SARS-CoV Urbani in 10-week-old BALB/c mouse lungs also resulted in mouse adapted strains after 15 or 25 2-3 day passages. Under near identical conditions in 1-year-old animals, only 5 passages were sufficient for mouse adaptation demonstrating the increased susceptibility of aged populations to lethal SARS-CoV infection (data not shown). Importantly, these viruses cause an ARDS phenotype reminiscent of acute human infections in aged animals, and pneumonia in young animals. Thus, age, genetic background of the host and immune status likely facilitate cross-species transmission and adaptation potential of zoonotic RNA viruses. Sequence analysis of young-mouse-adapted strains revealed mutations in common genes sets, notably nsp5, nsp9, the M glycoprotein genes, including changes within the RBD of S. Unique targets in icMA15 and MA25 included nsp3, nsp12 and nsp13. Using recombinant chimeras between icMA15 and wildtype SARS-CoV, mutations in nsp9 and the S glycoprotein were most important for eliciting fatal disease in young, but not aged animals, although these changes were still highly interdependent on the other mutations as well. In aged animals, however, smaller subsets (2 mutations) of mutations encoded by icMA15 genotype were capable of producing fatal disease. Interdependency of the allele sets was mostly lost as virtually all two or four set combinations tested were lethal and the S gene mutation set also resulted in significant morbidity with an ARDS pathogenic phenotype. These data are consistent with the idea that aged populations are more compliant hosts for rapid SARS-CoV adaptation to virulence.

Although the mutation sets in S were different in icMA15 and MA25, both

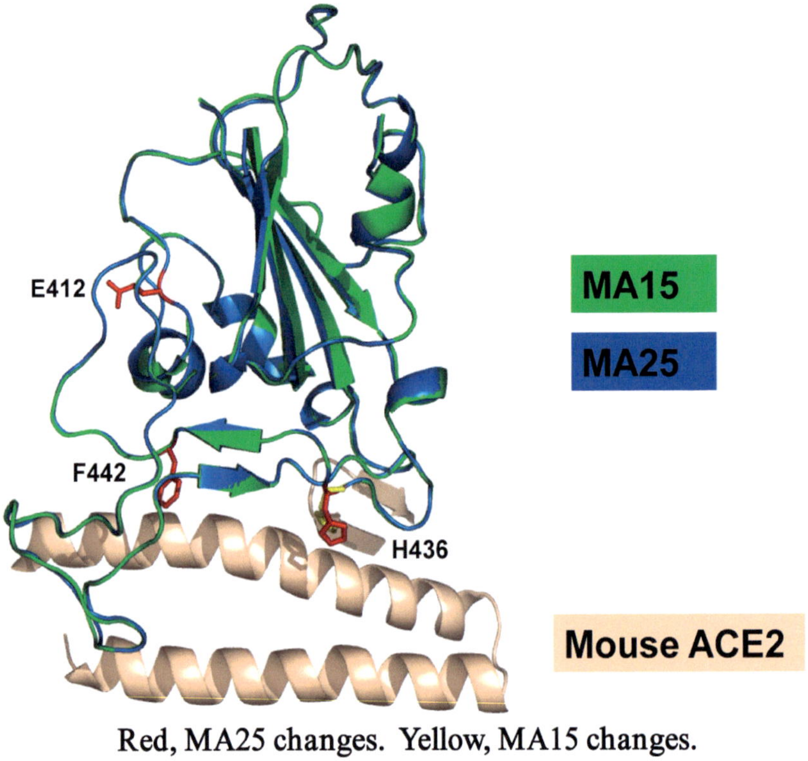

FIGURE J.2 Mouse adapted mutations in the S glycoprotein RBD-hACE interaction site.

sets of RBD mutation putatively enhanced S interaction with the mouse ACE2 receptor, likely increasing virus replication efficiency in the lung by 2-3 logs of titer (See above Figure. Mouse Adapted Mutations in the S Glycoprotein RBD-hACE Interaction Site). Other unique pathways to SARS-CoV S RBD mACE2 recognition have also been identified (data not shown), reiterating the plasticity of coronavirus RBD-receptor interaction networks. Thus, the *in vivo* models of SARS-CoV adaptation to the murine host reflected the key role that the S glycoprotein plays in mediating cross-species adaptation and pathogenesis. Notably, and in contrast to results seen with filoviruses, no evolution was noted in the six different antagonists of innate immunity. It is noteworthy that each *in vivo* adapation model selected for unique sets of mutations that enhanced virus replication efficiency, resulting in increased pathogenic outcomes. Although mutation sets targeted key genes essential for efficient virus-host interaction, the actual mutation sets were different, demonstrating the existence of multiple genetic pathways to increased virulence and the same pulmonary disease outcome. Moreover, host background (e.g., genetics, age, immune status, etc.) was a major selective force in the evolution of virulence, further complicating the development of universal laws governing cross-species transmission and virulence.

GENE REGULATION

Regulation of Bacterial Pathogenicity

If an organism possesses specialized gene products for its virulence, it must be able to utilize them when needed, but not squander its metabolic energy producing them aimlessly or risk having them detected by host defenses and neutralized prematurely. Consequently, regulation of virulence factor expression is an additional, yet essential, complication of a pathogenic microorganism's life. The host presents an array of conditions that are strikingly distinct from those of the outside environment, conditions that are not easily reproduced in the laboratory. In fact, laboratory culture conditions bias our understanding of microbial adaptation to natural environments. *V. cholerae*, for example, is thought to persist in brackish estuaries and other saline aquatic environments, sometimes associated with the chitinous exoskeleton of various marine organisms. Transition from this milieu to the contrasting environment of the human small intestinal lumen must be accompanied by substantial genetic regulatory events.

The microbial cell is relatively simple, yet it possesses the means to detect, often simultaneously, changes in temperature, ionic conditions, oxygen concentration, pH, and calcium, iron, and other metal concentrations that might appear to be subtle signals, but which are essential for the precise mobilization of virulence determinants. Similarly, environmental regulatory signals prepare the microorganism for its transition from an extracellular to an intracellular existence. For example, iron is a critical component of many cell metabolic processes; therefore, it is not surprising that animals rely on high-affinity iron-binding and storage proteins to deprive microorganisms from access to this nutrient, especially at the mucosal surface. In turn, most pathogens sense iron availability and induce or repress various iron acquisition systems accordingly. Indeed, many microorganisms possess toxins that are regulated by iron such that low iron concentrations trigger toxin biosynthesis. In addition, reversible regulation of virulence gene expression by temperature is a feature common to many pathogens. For example, *E. coli* may be deposited in feces and live for long periods of time under conditions of nutrient depletion and low temperature. However, it has learned to mobilize its colonization-specific genes when it is returned to the warm mammalian body. The regulatory machinery used to accomplish this is an important feature of many pathogens, including *Yersinia pestis* and *B. anthracis*, both of which exist in multiple niches within and outside the mammalian host.

The number of well-characterized virulence regulatory systems is rapidly increasing, in part because of the development of rapid methods for screening gene expression on a genome-wide basis (e.g., with the use of DNA microarrays). At the same time, relatively little is known about both the specific environmental signals to which these systems respond and the exact role of these responses in the course of human infection. One common mechanism for bacterial transduction of environmental signals involves two-component regulatory systems that act on gene expression, usually at the transcriptional level. Such systems make use of similar pairs of proteins; one protein of the pair spans the cytoplasmic membrane, contains a transmitter domain, and may act as a sensor of environmental stimuli, whereas the other is a cytoplasmic protein ("response regulator") with a receiver domain that regulates responsive genes or proteins. These regulatory systems are common in both pathogens and non-pathogens so their detection by sequence analysis cannot be employed as a reliable predictor of pathogen vs. non-pathogen.

The coordinated control of pathogenicity incorporates the important concept of a *regulon*. A regulon is a group of operons and/or individual genes controlled by a common regulator, usually a protein activator or repressor. A regulon provides a means by which many genes can respond in concert to a particular stimulus. At other times the same genes may respond independently to other signals. Global regulatory networks are a common feature of microbial virulence as well as basic microbial physiology and therefore their sequences, while often essential for a pathogen, are also not reliable predictors of virulence. The apparent complexity of virulence regulation in a single microbial pathogen is magnified by the coexistence of multiple interacting ("cross-talking") systems and by regulons within regulons.

Proper presentation of certain virulence-associated gene products on the microbial surface is now recognized to be as important to pathogenicity as the initial expression of these genes. Presentation entails export pathways, association with other periplasmic or surface factors, macromolecular assembly at the surface (for some), as well as regulation of the export components themselves. Among bacterial pathogens and non-pathogens alike that inhabit the human body, shared homology is apparent among families of proteins involved in these processes. Folding, transport, and assembly of specific proteins enable a microorganism to present a specific array of surface molecules necessary for eukaryotic cell tropism, intoxication, or entry. The precise configuration of a number of microbial surface molecules might be viewed as a cooperative "attack complex," a property not found in any of the individual components. However, remarkably similar complexes are used by harmless microorganisms in the colonization of mucosal surfaces. In one case, the harmless microorganism remains attached to the host surface and causes no harm while, in the other case, the organism attaches to the mucosal surface and breaches this barrier to cause infection and disease.

Pathogenic microorganisms have developed many different mechanisms by which to regulate expression of virulence factors. For organisms with multiple hosts, such as *Y. pestis*, differential expression of virulence factors is required for adaptation and survival within each host environment. In the case of commensal organisms that cause infection opportunistically, such as *Staphylococcus aureus*, control of virulence factor expression is niche-dependent.

The lifestyle of *Y. pestis* requires survival of the organism in the flea host and the mammalian host. The primary reservoir for *Y. pestis* is rodents. Rodent-to-rodent transmission can occur, which can lead to epizootic plague. Fleas become infected by feeding on rodents that are bacteremic as a result of *Y. pestis* infection. The bacteria multiply rapidly in the flea midgut, and cause blockage of the proventriculus. Transmission of *Y. pestis* from flea to human occurs as the flea bites and attempts to feed on the human. Because the proventriculus is blocked, the flea regurgitates the *Y. pestis* into the bite, which, in turn, may lead to human infection. The virulence factors required for each host environment are regulated primarily by the temperature of the host. LcrF and Caf1R are positive regulators of virulence factors that are expressed in the mammalian host (Reviewed by Konkel and Tilly 2000). LcrF regulates expression of ~50 genes that encode the Yops (*Yersinia* outer proteins); the Yop effector proteins are secreted via a type III secretion system that is part of the Yop family. In addition to being controlled by temperature, *lcrF* is regulated post-transcriptionally because the ribosome binding site of the *lcrF* transcript is sequestered in a stem-loop structure at low temperature. When the host temperature increases (i.e., when the bacteria enter the mammalian host), the stem-loop unfolds and permits binding of the ribosome to promote translation of *lcrF*. Caf1R regulates expression of the *Y. pestis* capsule genes. The capsule provides protection from phagocytosis of *Y. pestis*; such protection is critical to survival of the organism in the mammalian bloodstream. In contrast, the *hms* locus is induced at low temperature, when *Y. pestis* is in the flea. The *hms* locus encodes genes whose products are required for hemin storage. Inactivation of the *hms* genes results in *Y. pestis* strains that are unable to block the flea proventriculus, but, as anticipated, has no effect on pathogenesis in mammalian hosts because the genes are expressed only at low temperature (in the flea).

In conclusion, the inherent pathogenicity of a microorganism can be altered by gene additions or losses or through genetic regulation of essential genes for virulence.

Appendix K

Interactions of Infectious Agents with the Host

PLANT PATHOGENS

Examples are given by the major evolutionary mechanisms by which plant pathogens have emerged as threats to agricultural ecosystems. These mechanisms take place over variable time scales ranging from short to hundreds or thousands of years.

Host-Tracking

Host-tracking refers to a co-evolution of a pathogen with its host during the process of host domestication, which includes the formation of a specific agro-ecological system. Host-tracking includes the selection and cultivation of desirable host genotypes, processes that simultaneously select for pathogen genotypes adapted to the selected individuals and the agro-ecological conditions in which the process occurred. The process can take seven to twelve thousand years and pathogen and host share the same center of origin. A documented example is *Mycosphaerella graminicola* on wheat, (Stukenbrock, Banke et al. 2007). The emergence of this pathogen causing the Septoria tritici leaf blotch disease on wheat was studied by genealogical and model-based coalescent approaches on seven selected genes from 184 isolates (Stukenbrock, Banke et al. 2007). Two related but genetically differentiated wild grass-infecting populations of *M. graminicola* named S1 and S2, and adapted to wheat, were identified on wild grasses collected in northwest Iran. The S1 and S2 populations were encountered on three different weedy grass species growing in the proximity of fields cultivated with wheat. The analysis indicated that the split between the most closely related wild grass infecting population S1 and *M. graminicola* occurred approximately 10,000 to 12,000 years ago, which coincides with the time that wheat was domesticated, suggesting a co-speciation of the host and pathogen. Similarly, another fungal pathogen *Magnaporte oryzae*, the causal agent of rice blast disease, underwent a host speciation due to the loss of the avirulence gene *AVR-Co39* that was frequent in haplotypes from other hosts. The genetic isolation and divergence of an epidemic lineage from other populations on grasses may have resulted from rapid clonal propagation and strong selection mediated by the new domesticated host and its associated agro-ecosystem. The subsequent intensification and dissemination of this crop favored the propagation and global dispersal of clones of the rice-infecting blast pathogen (Couch, Fudal et al. 2005).

Host Shift

Host shift is a process in which a new pathogen emerges by adaptation to a new host that is a close relative of the former host (e.g., shifting from a wild crop to the new domesticated selection or variety of the crop). The process takes from less than 500 to 7,000 years, and the pathogen and host do not always originate in the same center of origin (Stukenbrock and McDonald 2008). An example is the shift of *Rhynchosporium secalis* from wild grasses to barley and rye. This was an abrupt evolutionary change that took approximately 2,500-5,000 years, much later than the domestication of barley and rye. *R. secalis* causes scald diseases and infects barley (*Hordeum vulgare*), rye (*Secale cereale*), and other grasses. A RFLP allelic diversity analysis of 1,366 geographical isolates indicated that the center of diversity of this pathogen did not coincide with the center of origin of barley, similar to the pattern found for the avirulence gene *NIP1* (O.24). *NIP1* has a dual function as both an elicitor of plant defense and as a toxin-encoding gene (O.25). *NIP1* is often deleted (O.26) because its role as elicitor for plant defense drives this gene under positive diversifying selection. Further analysis of *R. secalis* revealed phylogenetic relationships between different host-related lineages (O.27). These analyses confirmed a later emergence of the scald pathogen, most likely between 1,200-3,600 years after the introduction of the barley agro-ecosystem into northern Europe.

Host Jump

Host jump describes a process through which a new pathogen emerges in a host species that is genetically distant from the original plant host (e.g., from another class or order). In this case, the geographical origin of the host does not always correspond with the geographical origin of the pathogen as observed in the host shift process. An example is the emergence of *Magnaporthe oryzae*, which also fits into a host-tracking co-evolutionary scenario with rice; however, the emergence of a rice-infecting lineage also involved a number of host shifts from Setaria millet to rice (Couch, Fudal et al. 2005). The close proximity between crop plants can facilitate a host jump of pathogens associated with either one of the species. We will consider the evolutionary relationship between *M. oryzae* haplotypes on rice and on weeds of rice. The more ancestral haplotypes originated from rice, whereas haplotypes from weeds of rice were found at the tips of a haplotype network. After additional host jumps that occurred to common weeds of rice, including cutgrass (*Leersia hexandra*) and torpedo grass (*Panicum repens*), the host specialization of the rice-infecting lineage occurred (Couch, Fudal et al. 2005).

HOST SPECIES TROPISM

Variola and Monkeypox Viruses

What we know concerning variola and monkeypox virus virulence genes involved in host species tropism comes by analogy to well-characterized orthopoxvirus orthologues. Different poxviruses encode a different pattern of virulence genes that is the basis for their unique host biology. The function of a large number of these virulence genes is to ensure the infected cell or neighboring cells are metabolically active and capable of efficient production of progeny virus. Virulence genes encoded by monkeypox and variola viruses are listed for convenience as targeting five pathways:

- inhibitors of apoptosis, an early cellular protective response that eliminates virus-infected cells and limits virus replication;
- inhibitors of pathogen recognition receptor (PRR) pathways that are triggered by pathogen associated molecular patterns (PAMPs) such as dsRNA and DNA;
- inhibitors of the interferon response, which induces a large number of unique antiviral molecules;
- modulators of the ubiquitin ligase system that regulates a large number of intracellular processes
- modulators of cell cycle and processes associated with transcription, DNA replication, and protein synthesis.

Genome comparisons identified 12 virulence genes that differed between variola and monkeypox viruses. The majority of these genes affect cell processes (i.e., apoptosis blockers, 2 genes and PRR and IFN responses, 7 genes). These differences are hypothesized to explain the dramatic difference in the animal species that act as reservoir or incidental hosts for the viruses. Monkeypox virus has a broad host range and several animal species may act as reservoir hosts in nature. In addition, field studies conducted in the lowland tropical forests of the Congo Basin and West Africa revealed that monkeypox virus can infect many animal species, including squirrels (*Funisciurius* spp. and *Heliosciurus* spp.)

and non-human primates (such as *Cercopithecus* spp.; Parker S). Variola virus, on the other hand, has human as the sole reservoir species and fails to cause experimental disease in standard adult, small animal, and non-human primate models under physiological conditions.

Franciscella Tularensis

Tularemia is a zoonotic disease caused by one of several subspecies of *Francisella tularensis*. *F. tularensis* subsp *tularensis* and subsp *holarctica* are most commonly associated with disease in humans. Outbreaks of disease in humans commonly occur during disease cycles in rodents and lagomorphs, mostly as a result of transmission from one mammal to another by one of a number of arthropod vectors. *F. tularensis* is a facultative intracellular bacterium that can infect humans via the skin (ulceroglandular), the conjunctiva (oculoglandular), the mouth (oropharyngeal or gastrointestinal), or the airway (pneumonic). Macrophages are the primary target cell for *F. tularensis*, which is taken up by asymmetric pseudopod loops formed by the macrophage in a complement-dependent manner. Intracellular survival of *F. tularensis* is a complicated process that involves, but is likely not limited to, genes encoded on the large *Francisella* pathogenicity island. Genomic and proteomic analysis of *F. tularensis* spp. identified a large number of hypothetical genes/proteins with no homology to known genes and proteins. Thus, *F. tularensis* spp. provide ideal examples of organisms for which knowledge of genomic sequence does not allow for prediction of virulence.

Rickettsia Species

The *Rickettsia* genus contains at least 18 species, all of which are flea- or tick-borne bacteria (Gillespie, Ammerman et al. 2009). *R. prowazekii* and *R. typhi* are the etiological agents of typhus and murine typhus, respectively, and *R. rickettsii* is the causative agent of Rocky Mountain Spotted Fever. *R. prowazekii* is transmitted from louse to mammalian host in the feces of the louse; the rickettsia are deposited onto the host skin by the louse and subsequently enter the bloodstream of the host via scarification of bites. Naïve lice become infected through feeding of the mammalian host, thus renewing the life cycle of the rickettsia. *R. typhi* are transmitted to rodents, primarily rats, by a similar mechanism except that the arthropod vector is the flea. An uncommon sequela to typhus is the development of a recrudescent illness, known as Brill-Zinsser disease, during which *R. prowazekii* can sequester itself within the host for months to years. Because infection of a mammalian host results from death of the louse, Brill-Zinsser disease may represent a mechanism by which *R. prowazekii* can be maintained and transmitted for many years following the initial infection. *R. rickettsii*, on the other hand, is transmitted to its mammalian host through the bite of a tick; the tick can infect multiple hosts because it does not die as a result of the *R. rickettsii* infection. Little information is available about the rickettsial genes that are required for entry into and replication within the arthropod vector of any of the known *Rickettsia* species. The genomes of the typhus-causing rickettsia and *R. felis* (another insect-borne *Rickettsia* species) are surprisingly different given that they share similar arthropod hosts and cause similar diseases (of varying intensity) in their mammalian hosts. Of note, only two open reading frames are shared specifically by the insect-borne *Rickettsia* species (*R. prowazekii*, *R. typhi*, and *R. felis*); these genes are linked on the chromosome and are thought to have been acquired by lateral gene transfer, which is a newly discovered phenomenon in the rickettsial species (Gillespie, Ammerman et al. 2009). However, the function of the open reading frames is unknown. Thus, once again, knowledge of the genome sequence of a group of related organisms does not allow us to predict the pathogenicity of any of the organisms.

HOST INNATE RESPONSES AND PATHOGEN-ENCODED COUNTERMEASURES

Microbial pathogens have evolved complex and efficient methods to overcome both innate and adaptive immune host responses to infection. Here we provide a few examples from the large number of the diverse approaches that viruses and bacteria have evolved to evade and subvert key, innate antimicrobial responses in animals and plants.

Plant Innate Defense System

Plants have evolved resistance (*R*) genes encoding proteins that confer resistance to specific pathogens. The plant pathogen molecule that specifically elicits R-protein-mediated responses is termed an avirulence (Avr) determinant. The Avr proteins are usually necessary for successful infections and are virulence factors in a susceptible host (Soosaar, Burch-Smith et al. 2005). For example, the *Arabidopsis thaliana RCY1* gene confers resistance to the Y strain of Cucumber mosaic virus (CMV), but not to the O strain. When the Y strain of CMV infects *RCY1*-containing plants, a defense response that restricts the virus to the infection site and prevents disease is initiated. The virus is an avirulent pathogen on these resistant plants and this is termed an incompatible interaction (Soosaar, Burch-Smith et al. 2005). *A. thaliana* encodes a second *R* gene, *HRT*, which confers a hypersensitive response (HR) to *Turnip crinkle virus* (TCV). Both *RCY1* and *HRT* genes are allelic and encode proteins that share 91 percent similarity but confer resistance to unrelated viruses: CMV, a cucumovirus and TCV, a carmovirus, respectively. Another example in *Nicotiana glutinosa* is the *N* gene which confers resistance to *Tobacco mosaic virus* (TMV, a Tobamovirus) (Soosaar,

Burch-Smith et al. 2005; Bent and Mackey 2007). In order to establish a rapid and productive infection, a plant virus must enter the plant cell with defense proteins or immediately synthesize them. Examples include the following:

- The P1-HcPro encoded by *Turnip mosaic virus* (TuMV), a virus with a RNA genome, interferes with the miRNA-controlled development pathways that share components with the antiviral RNA-silencing pathway. This interference acts as a viral counter-defense mechanism that enables systemic infection by TuMV (Dunoyer, Lecellier et al. 2004);
- The p19 proteins from a number of tombusviruses including *Tomato bushy stunt virus*, *Cymbidium ringspot virus*, and *Carnation Italian ringspot virus* allow a high accumulation of viral RNAs and also are responsible for TBSV pathogenesis (Qiu, Park et al. 2007).
- The CP (capsid protein), p20, and p23 proteins of Citrus tristeza virus each have an unique suppressor of RNA silencing activity (Lu, Folimonov et al. 2004).
- The 2b protein encoded by CMV performs as a suppressor of RNA silencing and a pathogenicity determinant (Anandalakshmi, Pruss et al. 1998; Brigneti, Voinnet et al. 1998).
- The p25 of *Potato virus X* which blocks the host silencing signal from spreading to other cells (Voinnet, Lederer et al. 2000).
- *Pseudomonas syringae* AvrPto and AvrPtoB act upstream of the MAP kinase signaling to suppress transcription of a few transcripts induced by flagellin via PRR FLS2.
- Other bacteria with avirulence genes are: *Pseudomonas syringae* pv. *glycinea* (*avrA*, *avrB* and *avrC*); *Pseudomonas syringae* pv. *Tomato* (*avrD*, *avrRpt2*, and *avrPto*); *Xanthomona campestris* pv. *Vesicatoria* (*avrBs1*, *avrBs2*); and *Xanthomonas campestris* pv. *raphani* (*avrXca*) (Vivian and Gibbon 1997)
- A number of fungal virulence genes have also been discovered: *Nectria haematococca* (*PEP* and *PDA*); tomato leaf mold fungus *Cladosporium fulvum* (syn. *Passalora fulva* (*Avr*s and *Ecp*s); and *Magnaporte oryzae* (*AVR1-CO39*) (van der Does and Rep 2007).

Pattern-Recognition Receptor Signaling

The plant and animal germ-line encoded PRRs of the innate immune system sense through pathogen-associated molecular patterns (PAMPs) the presence of a bacterial, viral or fungal infection (Medzhitov R). One type of PRR, the toll-like receptor (TLR) family, recognizes a vast array of microbial molecules, including lipopolysaccharide (TLR4), bacterial flagellin (TLR5), viral double-stranded RNA (TLR3), and bacterial and viral DNA (TLR9). All PRRs initiate signaling pathways that converge at the activation of the transcription factors IRF3, IRF7, and/or NF-κB, which leads to the expression of IFN-β and the IFN-stimulated genes. Some PRRs also instruct the adaptive immune system, thereby orchestrating an optimal response against the particular pathogen.

Viruses more so than bacteria directly interfere with PRR signaling. The importance of this pathway in the control of virus replication is underscored by the identification of greater than 19 inhibitors of PRR signaling encoded by protypic viruses from 9 virus families (Bowie and Unterholzner 2008). Some viruses interfere in the PRR signaling pathway at multiple points as mentioned previously for variola virus (i.e., Cop-M2, Cop-K1, Bsh-D7, Cop-N1 and Cop-A46). Similarly, hepatitis C virus NS3-4A protein inhibits TLR3 signaling though the degradation of TRIF, and the NS5A protein inhibits the activity of IFN-inducible dsRNA-dependent protein kinase, 2',5'-oligoadenylate synthase, and myeloid differentiation primary-response gene 88. In addition, a single step in the PRR pathway can be targeted by convergent evolution by a number of viruses. Ebola virus VP35, vaccinia virus E3L, influenza virus A NS1 and reovirus σ3 all sequester double-stranded RNA that prevents activation of RIGI and MDA5. Evidence of bacterial pathogens directly interfering with TLR signaling is limited; however there is at least one example of downstream modulation of PRR responses. LcrV is encoded on pYV virulence plasmid common to *Y. pestis*, *Y. pseudotuberculosis*, and *Y. enterocolitica*. Like many virulence factors, LcrV is described to be involved in several functions including regulation of production of Yops and the type III protein secretion system that translocates Yops into host cells. In addition, LcrV is an immunomodulator involved in TNF-α and IFN-γ down-regulation and IL-10 induction through interaction with cell surface CD14/TLR2 (Sing, Reithmeier-Rost et al. 2005). Plant pathogenic bacteria also target plant PRR using mechanisms both unique to plants and conserved in plants and animals. One such conserved mechanism is the delivery of type III effector proteins via a type III secretion system present in *P. syringae* and other Gram-negative pathogens. There is a great diversity of effectors both within and among bacterial species based on sequence level comparisons; over 30 effectors are likely to be delivered by *P. syringae* pathovar *tomato* (Chang, Urbach et al. 2005). One effector, AvrPto, suppresses signaling from the plant surface PRR FLS2 that senses the conserved flg peptide of flagellin (Abramovitch, Anderson et al. 2006).

Complement

The complement system in mammals consists of more than 35 soluble proteins and receptors that play a key role in innate and adaptive immunity. Complement is activated through three different pathways: alternative, lectin and classical. In innate immunity, the functions mediated by complement activation products include phagocytosis and cytolysis of pathogens, solubilization of immune complexes, and inflammation (Walport 2001a, 2001b). The complement system is present in invertebrate species and is at least 600 million years old, which explains the varied mechanisms employed by microorganisms to block or subvert its action.

To date bacteria, viruses, fungi, and parasites have been reported to encode 38, 10, 3, and 8 distinct proteins, respectively, which target the complement pathway (Lambris, Ricklin et al. 2008). The activity of these proteins falls into three classes: (1) recruitment or mimicking of complement regulators; (2) modulation or inhibition of complement protein by direct interactions; and (3) inactivation of complement regulators by enzymatic degradation. In addition, bacteria and viruses employ passive features to subvert complement function. The surfaces of certain enveloped viruses such as orthopoxviruses contain host-derived regulators of complement activation, decay-accelerating factor and membrane cofactor protein which block lysis (Moulton, Atkinson et al. 2008). The cell walls of the Gram-positive bacteria such as *S. pneumoniae* and *S. aureus* are resistant to membrane-attack complex formation (Joiner, Brown et al. 1983). *M. tuberculosis* employs a C2a-dependent entry pathway that results in surface deposition of C3b. The opsonized bacteria are taken up into the macrophage via the complement receptor CR3 where they replicate within phagosomes (Schorey, Carroll et al. 1997).

Inflammatory/Immune Cytokines

Cytokines and chemokines play crucial roles in inducing the migration of inflammatory/immune cells to areas of infection, in antimicrobial defense, and in the orchestration of the adaptive immune response. As such, cytokine/chemokine signaling pathways are key targets of microbial evasion and subversion mechanisms that act inside the cell, at the plasma membrane, and outside the cell.

Microbial pathogens block or subvert the intracellular production and/or intracellular signaling of cytokines and chemokines by a number of mechanisms. A common target for microbial pathogens is the transcription factor NF-κB. Poxviruses encode 15 unique proteins that inhibit different steps in the signaling pathways leading to NF-κB activation, and an additional 5 proteins that uniquely and directly interact with the NF-κB complex (Mohamed, Rahman et al. 2009). *S. flexneri* type III secretory system effector OspG and rotavirus NSP1 protein inactivate the cellular E3 ubiquitin ligase complex $SCF\beta^{-TrcP}$, which is required for degradation of IκBα (Kim, Lenzen et al. 2005; Graff, Ettayebi et al. 2009). YopJ (*Y. pseudotuberculosis*) inhibits MAPK (mitogen-activated protein kinase) and NF-κB activity. The NF-κB inhibitory activity is likely mediated by yet another unique mechanism as YopJ has homology with the cysteine proteases of the "ubiquitin-like protease" family, and the substrates for YopJ were shown to be highly conserved ubiquitin-like molecules (e.g., SUMO-1) (Orth K). The YopJ family contains proteins found in other microorganisms with intimate relationships with eukaryotes, both animals and plants (e.g., adenovirus, adenovirus protease; *S. typhimurium*, AVrA; plant pathogen, *Xanthomonas campestris*, AvrBsT). Importantly, the predicted catalytic triad of YopJ and AvrBsT were required for inhibition of MAPK and NF-κB signaling in animal cells and for the induction of localized cell death in plants, respectively (Neish 2004).

Viruses also modify the activity of cytokines/chemokines at the plasma membrane prior to engagement of the receptor. For example, herpesviruses and poxviruses collectively encode over 40 viral members of the seven transmembrane–spanning G protein-coupled chemokine receptor superfamily (Sodhi, Montaner et al. 2004). Other classes of viral virulence genes interfere with cell surface binding of cytokines. The poxvirus IFN-α/βBP can bind type I IFNs in the extracellular milieu, and can also bind back to the surface of infected or uninfected cells to act as a decoy receptor, preventing the binding of type I IFNs to cellular receptors and the induction of an antiviral response (Xu, RH).

Finally, cytokines/chemokines are sequestered or destroyed outside the cells (Alcami 2003). Some pathogens sequester cytokines by soluble, virus homologs of host receptors or binding proteins [e.g., orthopoxviruses have binding proteins for IL-1β, IL-18, TNF, CD30, type I and type II IFNs (Alcami 2003)]. Certain viruses downregulate the cellular synthesis of pro-inflammatory cytokines and/or other antiviral functions of innate cells by expressing viral homologues of cytokines IL-10 (Epstein Barr Virus, BCRF1; orf virus, vIL-10), IL-17 (herpesvirus saimiri, ORF 13), IL-6 (Karposi's sarcoma-associated herpesvirus, K2), and semaphorin (asinine herpesvirus, SEMA; vaccinia virus A39). Chemokines are also targeted by pathogens in order to disrupt the chemokine gradient that diminishes or blocks the migration of inflammatory cells to a focus of infection in tissue. The chemokine gradient is disrupted by sequestering or destroying the relevant chemokine(s) [e.g., poxviruses, binding proteins for CC-chemokines; herpesviruses, binding proteins for CC-, CXC-, C-, and CX_3C-chemokines; Alcami A)] or by the release of pathogen-encoded antagonists or agonists (e.g., poxviruses, antagonist, for CC- and CXC; herpesviruses, agonists for CC-, CCR8, CCR4, and CXCR2; antagonists for C-, CC, CXC, and CX_3C-; Alcami A). Another example of this strategy can be found with *Streptococcus pyogenes*. *S. pyogenes* has several mechanisms to modulate neutrophil-mediated antibacterial activity, including the targeted degradation of chemoattractant molecules C5a by SCPA (Ji Y) and IL-8, GCP-2 and GROα by spyCEP(Sumby, Zhang et al. 2008).

Inflammatory Cell Response

Natural killer cells, macrophages, neutrophils, and dendritic cells are key inflammatory cells of the innate response. The latter three cell types are important in the control of both bacterial and viral infections through the release of cytokines/chemokines and microbicidal factors, and phagocytosis of the pathogen. Due to the 1-3 micron size of bacteria, phagocytosis is an effective mechanism for bacterial clearance, and has driven the evolution of microbial countermeasures to a number of steps in the phagocytosis pathway.

Translocated bacterial effector proteins can kill cells through necrosis (i.e., toxins) or by inducing apoptosis or inhibiting anti-apoptotic signaling, which can prevent necrotic release of proinflammatory signals (e.g., *Shigella flexneri*, IpaB (Zychlinsky, Kenny et al. 1994); *Yersinia pseudotuberculosis*, YopJ (Orth K). Other effectors inhibit uptake of microorganisms by disrupting the host cell cytoskeleton (e.g., *Y. pseudotuberculosis*, YopE and YopH (Fallman, Andersson et al. 1995; Black and Bliska 2000). Certain microorganisms are able to escape from the phagosome and/or block the phagosome-lysosome fusion. *L. monocytogenes* is an example of a microbe that efficiently escapes the phagosome and replicates in the cytosol. The phagosomal membrane is disrupted by the action of listeriolysin O and two membrane-active phospholipase C enzymes (e.g., phosphoinositol-specific phospholipase C and broad-range phospholipase C encoded by genes *plcA* and *plcB*, respectively; (Flannagan, Cosio et al. 2009). *M. tuberculosis* is an example of a microbe that arrests phagosomal maturation. An array of factors, including the lipids phosphatidylinositol mannoside and lipoarabinomannan, as well as phosphatidylinositol-3-phosphate phosphatase SapM, prevent the transition of early phagosomes to the late and phagolysosomal stages (Flannagan, Cosio et al. 2009).

SELECTION PRESSURE

Borrelia Relapsing Fever

Tick-borne relapsing fever is caused by *Borrelia hermsii*, a spirochete that infects the *Ornithodoros hermsi* tick. *O. hermsi* is a fast-feeding tick that transmits *B. hermsii* during a blood meal. Characterization of the bacterial surface proteins during different stages of its life cycle revealed that the spirochetes alter their surface in response to their environment (Schwan and Piesman 2002). When the spirochetes are in the gastrointestinal tract of the tick, they express an Outer Surface Protein (Osp) identical to the Osp expressed in their most recent previous mammalian host. The same is true during the first spirochetemic phase of mammalian infection. However, the spirochete undergoes antigenic variation in the host as a mechanism by which to evade the host immune response. As many as 30 different versions of a single Variable Major Protein (Vmp) of *B. hermsii* have been identified following outgrowth from a single starting cell in the presence of selective pressure. Each round of spirochetemia (relapsing fever) occurs as a result of antigenic variation of the Vmp so that the pathogen can evade the host defenses mounted by the host to overcome the previous fever episode.

RNA Viruses

As noted earlier, filoviruses and highly pathogenic coronaviruses target a variety of host pathways to enhance virus cross-species transmission, host range and virulence. Among filoviruses, these seem to be mediated by mutation driven processes, while coronaviruses utilize a mixture of recombination and mutation driven pathways to evolve and/or acquire new gene functions. Virulence genes encoded by filoviruses and coronaviruses are listed for convenience as targeting several unique and common pathways.

- Virus-receptor interactions to promote cross species transmission (SARS-CoV S; Filoviruses: role of GP-mediated host range less clear)
- Proapoptotic genes that contribute to virus induced cell killing and pathogenesis (SARS-CoV-ORF3a/b, ORF6, ORF7a/b, ORF8, S and M glycoproteins; Filoviruses: GP/sGP);
- Inhibitors of the interferon response, which induces a large number of unique antiviral molecules; (SARS-CoV: nsp1, nsp3(PLP), nsp7, nsp15, ORF3b, ORF6 and N; Ebola Virus: VP24 and VP35)
- Inhibitors of NF-kB signaling machinery (SARS-CoV: nsp1, PLP)
- Inhibitors of nuclear import machinery (SARS-CoV-ORF6; Ebola-VP24)
- Modulators of the ubiquitin ligase system or SUMO modification machinery that regulate a large number of intracellular process; (SARS-CoV: nsp3(PLP) deubiquitinase activity; Ebola VP35-causes increased SUMOylation of IRF7)
- Modulators of cell cycle and processes associated with transcription, DNA replication and protein synthesis (SARS-CoV nsp1)
- Immunosuppression of adaptive immunity (Ebola/Marburg: GP immunosuppressive motif; SARS-CoV: lymphopenia and thrombocytopenia-genetic mechanisms unknown)

IMMUNODEFICIENCY

HIV Infection

The classical example of a microbe that suppresses the immune system is the Human Immunodeficiency Virus (HIV-1). Over time after infection with the HIV-1 virus, an untreated person's helper cells become depleted. If he/she was previously silently infected with *Mycobacterium tuberculosis* (the agent of tuberculosis or TB), as is an estimated one-third of the world's population, that individual will no longer be able to contain the replication of the TB organism. Indeed, in Africa, HIV-1-infected people who develop AIDS (Acquired Immune Deficiency Syndrome) as a consequence of a dearth of T-helper cells frequently die of tuberculosis. Conversely, in most immunologically normal people who are infected with *M. tuberculosis*, the bacterium remains walled off in granulomas by a T-helper-cell-dependent host adaptive immune response called cellular immunity.

Organ Transplantation

For organ transplant recipients, the risk of infection and the type of infection are functions of the degree of immunodeficiency (i.e., type of immunosuppressive regimen) and the use of preventive prophylaxis. Organ transplantation for end-stage organ failure is an effective therapy, but is limited by the availability of donor organs. In 2005, 66,000 kidney transplants, 21,000 liver transplants and 6,000 heart transplants were carried out worldwide (WHO volumes/85/12/06-039370). The use of increasingly potent immunosuppressive agents has reduced the incidence of rejection, but increased the patient's susceptibility to opportunistic infections. The sources of the infections are from the donor organ, the recipient (preexisting condition), the community or the hospital environment. Although a large number of infectious agents are capable of infecting transplant patients, the majority of infections are caused by the following pathogens: cytomegalovirus, Epstein-Barr virus, adenoviruses, polyomaviruses BK and JC, and *Pneumocystis* and various fungal species (Fishman 2007). One study documented fungal infections as a major cause of morbidity and mortality with incidence rates ranging from 5 percent among recipients of kidney transplants to as high as 40 percent among recipients of liver transplants (Paya 1993).

MICROBIOME

Humans are the natural host to a myriad of microorganisms that assemble into complex, largely beneficial communities that outnumber human cells by ten-fold. The dominant forms of human-microbe interactions are those in which microorganisms benefit the host without causing harm (commensal relationships), and relationships in which both host and microorganism benefit (symbiotic or mutualistic relationships). Co-evolution, co-adaptation and co-dependency are features of our relationship with our indigenous microbiota.

Our microbiota is ancient and largely conserved in the general types of organisms that inhabit and persist within us for life. Yet, the microbiota in vertebrates is not only often host-specific, it is also compartmentalized to be niche-specific. For example, the gut microbiota and oral microbiota have quite distinct microbial inhabitants. Although possibly germ-free (gnotobiotic) before birth, humans develop a resident microbiota shortly after birth. In the neonatal period, the community assembly process is dynamic and is influenced by early environmental (in particular, maternal) exposures and stochastic effects. The composition of the indigenous microbiota evolves in a generally orderly fashion in response to diet and other environmental factors; it is influenced as well by a diverse human genetic background. The bacterial diversity in the human body is striking in its richness of distinct species and strains, but also noteworthy for the limited number of phyla commonly found in indigenous microbial communities. Of the more than 50 bacterial phyla in the environment, only four (*Firmicutes*, *Bacteroidetes*, *Actinobacteria*, and *Proteobacteria*) dominate human mucosal and cutaneous habitats, suggesting that strong selective forces have limited diversity over at least hundreds of thousands of years of co-evolution (8-16). Despite this stereotypic assembly process, within a single mammalian species, including *Homo sapiens*, each individual has a virtually unique microbiome; its composition and the phenotypes expressed affect as well as reflect the overall biological diversity of humans.

The human microbiome is the subject of intensive study, including the major international Human Microbiome Project (HMP). Because of advances in DNA sequencing technologies and improvements in bioinformatics, it has become possible to characterize the great diversity in the human microbiota. In 2007, the National Institutes of Health (NIH) launched the Human Microbiome Project (HMP) as one of its major roadmap initiatives. This major scientific endeavor has the following aims:

- Determine whether individuals share a core human microbiome.
- Understand whether changes in the human microbiome can be correlated with changes in human health.
- Develop the technological tools to support these goals.
- Address the ethical, legal, and social complications raised by human microbiome research.

The human microbiome project will add an enormous amount of additional microbial sequence to our already burgeoning databases. This will be invaluable as we continue to sort out the sequences that have real predictive value instead of being merely suggestive because of some degree of relative homology with a putative virulence factor of a Select Agent.

Appendix L

Near-Term Milestones for Consideration

The committee finds that a gene-sequence based *classification* system is feasible using current technologies. Below the committee provides near-term milestones that would enable such a system. While the system could have advantages over the current system, there are also potential negative consequences that require careful deliberation. It is not the role of our committee to recommend specific implementation plans, nor are we properly constituted to do so. Many of the issues and priorities that must be considered are beyond the scope of our charge. Therefore, the committee has presented these ideas for discussion and recommends caution when considering the development of a gene-sequence-based oversight system.

(a.) ***A sequence database with a Select Agent focus:*** *(presented in Chapter 4)*

(b.) ***The expanding sequence database of all biology:*** *(presented in Chapter 4)*

(c.) ***Define the Criteria for Select Agent Designation:*** *(presented in Chapter 4)*

(d.) ***Stratification or reduction of the Select Agent list:*** *(presented in Chapter 4)*

(e.) ***Develop a Centralized System:*** To be useful for unambiguous regulations, there would need to be a single agreed-upon classification system, not multiple competing ones developed by different research groups. This would mean a centralized funding plan that would have to balance the benefits of single source standardization by a single SA classification system team against the need for oversight and review to maintain quality and efficiency in the absence of peer competition.

(f.) ***Scientific Workgroups and Advisory Panel:*** A work group of scientific experts would be assembled for each Select Agent (or Select Agent group). These expert groups would evaluate the agents and identify a "minimal parts list" for each. This is a highly technical and necessary step in developing a sequence-based classification system. The "Content Workgroups" would also identify genes necessary (though not sufficient) for virulence, and other "sequences of concern" that should be monitored as part of the yellow flag system.[1] An additional benefit of such "Content Workgroups" is that participation in this undertaking could raise awareness of dual-use issues among researchers. Moreover, to include the top experts, these content workgroups would necessarily include international scientists, which may strengthen international engagement. These are major objectives of the National Strategy of for Countering Biological Threats,[2] as well as the NSABB.

In addition to the "Content Workgroups," a panel of scientific advisors would be established for assessment of the Select Agent list and the yellow flag sequences. As previously mentioned, advisors for the yellow flag system would be charged to review biological data and make determinations as to whether a sequence raises sufficient concern to merit a yellow flag, needs further study, or should be removed from the yellow flag list. This panel would offer advice regarding whether a sequence construct merits consideration for Select Agent designation. The panel would be expected to consider information provided by the "Content Workgroups," and would likely have joint members.

The same (or a second) panel of scientific advisors would be charged with determining if biological criteria have been met to warrant designation of a pathogen or toxin as a Select Agent. In this capacity, the advisory panel would also work with stakeholders from the security community and government agencies, and therefore, the scientific advisors should be represented on (or function as a subcom-

[1] The committee stresses that this should not be confused with prediction. It would be a mistake to assume that these genes would function in a genome backbone independent manner, or to apply this information to designate a "sequence of concern" as a Select Agent. The purpose is to classify genomes as "complete" and subject to the SAR, and to identify "sequences of concern" that are worth monitoring.

[2] "The objectives of our Strategy are: . . . Reinforce norms of safe and responsible conduct: Activities that should be taken to reinforce a culture of responsibility, awareness, and vigilance among all who utilize and benefit from the life sciences to ensure that they are not diverted to harmful purposes. Transform the international dialogue on biological threats: Activities targeted to promote a robust and sustained discussion among all nations as to the evolving biological threat and identify mutually agreed steps to counter it."

mittee of) the NSABB, ISATAC, or the recently recommended Biological Select Agents and Toxins Advisory Committee (BSATAC).[3]

(g.) ***Updating, improvement, and assessment:*** The sequence-based classification system and the yellow flag system, once established, would be updated on a regular basis. The "Content Workgroups" would reconvene on a regular basis to incorporate the latest scientific information into the parts list. Any particular parts list would only reflect the current state of scientific knowledge about each Select Agent. It would need to be subject to review and revision to stay current with the state of knowledge distinguishing Select Agents from other organisms. Likewise, the profile classification system would have to be updated over time, as new knowledge accrued that required newly discovered SA or non-SA variants to be classified. This updating process would resemble the ongoing curation of other profile library classification systems such as TIGRfams and Pfam, and would require a curation team.

Assessments of the yellow flag and Select Agent lists—carried out by, or in collaboration with, scientific advisory panels—would occur on a regular basis. There is a need for increased transparency about the procedures and criteria for moving agents on and off the list.[4] The updating, review and assessment cycles are consistent with

[3]2009 NRC report Responsible Research with Biological Select Agents and Toxins, "RECOMMENDATION 2: To provide continued engagement of stakeholders in oversight of the Select Agent Program, a Biological Select Agents and Toxins Advisory Committee (BSATAC) should be established. The members, who should be drawn from academic/research institutions and the private sector, should include microbiologists and other infectious disease researchers (including select agent researchers), directors of BSAT laboratories, and those with experience in biosecurity, animal care and use, compliance, biosafety, and operations. Representatives from the federal agencies with a responsibility for funding, conducting, or overseeing select agent research would serve in an ex officio capacity. Among the responsibilities of this advisory committee should be the following: Promulgate guidance on the implementation of the Select Agent Program; Facilitate exchange of information across institutions and sectors; Promote sharing of successful practices across institutions and sectors; Provide oversight for evaluation of the Select Agent Program; Provide advice on composition/stratification of the list of select agents and toxins; Convene regular meetings of key constituency groups; and Promote harmonization of regulatory policies and practices (NRC 2009b)."

[4]2009 NRC report Responsible Research with Biological Select Agents and Toxins, "RECOMMENDATION 2: To provide continued engagement of stakeholders in oversight of the Select Agent Program, a Biological Select Agents and Toxins Advisory Committee (BSATAC) should be established. The members, who should be drawn from academic/research institutions and the private sector, should include microbiologists and other infectious disease researchers (including select agent researchers), directors of BSAT laboratories, and those with experience in biosecurity, animal care and use, compliance, biosafety, and operations. Representatives from the federal agencies with a responsibility for funding, conducting, or overseeing select agent research would serve in an ex officio capacity. Among the responsibilities of this advisory committee should be the following: Promulgate guidance on the implementation of the Select Agent Program; Facilitate exchange of

this goal. Moreover, they are important components for a robust and effective gene-sequence-based oversight system for Select Agents and "sequences of concern." The periodic expert review, assessment, and update cycle could be coordinated well with the biennial review of the Select Agent and Toxin list, which is currently required by statute.[5]

(h.) ***Disclosure, transparency, review:*** Because it is automatic and software-based, the classification system could be made readily and transparently available on the Web, where it could be reviewed and challenged by scientists in the community, to be sure (for instance) that it was not inadvertently misclassifying non-Select Agents such as vaccines or attenuated research strains. This feedback would inform the assessment and curation processes, and facilitate engagement of the scientific community. Likewise, the yellow flag biosafety system should be accessible and open for information sharing.

Should a gene-sequence-based system be developed? We have concluded that a gene-sequence-based classification system *could* be developed. We have not addressed whether such a system *should* be developed. Here are some issues for consideration:

a. ***Dual-Use Issue:*** As discussed, the development of a *prediction*-based oversight system would raise serious dual-use concerns. Providing an accurate mechanism for evaluating the threat posed by a synthetic genome sequence is equivalent to enabling the design and optimization of a bioweapon. It is a major goal of biology to predict phenotype from genotype, and to improve public health by understanding pathogenicity. However, it does not seem wise to make *special* plans for an advanced effort in predicting the sequence of would-be bioweapons.[6]

---

information across institutions and sectors; Promote sharing of successful practices across institutions and sectors; Provide oversight for evaluation of the Select Agent Program; Provide advice on composition/stratification of the list of select agents and toxins; Convene regular meetings of key constituency groups; and Promote harmonization of regulatory policies and practices (NRC 2009b)."

[5]"(42 U.S.C. 262a) . . . The Bioterrorism Preparedness Act requires that the HHS Secretary review and republish the list of select agents and toxins on at least a biennial basis." "(7 U.S.C. 8401) . . . The USDA Secretary is also required to conduct a biennial review of the USDA select agents and toxins list" (DHHS (2005). 42 CFR 72 and 73 and 42 CFR Part 1003: Possession, Use, and Transfer of Select Agents and Toxins; Final Rule, Federal Register. **70:** 12294-13325.

[6]If infallible prediction of organisms with Select Agent properties from genome sequence were feasible, or became feasible, it is unclear how that prediction would be useful in the context of Select Agents Regulation. Suppose we *could* predict the virulence, transmissibility, ease of growth, ease of dispersion, and environmental stability of a microorganism or virus from its sequence. Imagine we have a black box, a computer program that we can input a genome sequence to, and without error the black box reports whether the sequence corresponds to an organism that has

A narrow focus on milestones solely to be able to predict what makes an agent a threat to security may be a distortion of priorities in biology. Once biology in general approaches the goal of determining function from sequence, then it would be an appropriate time to consider a predictive oversight system to accurately identify Select Agent status from a novel genome sequence. This time may not come for decades, and may be more than a century away.

A classification system differs from a prediction system and would not directly enable the optimization of a pathogen. However, for a classification system to be usefully implemented, information must be shared. Listing the "parts" of a Select Agent and identifying other "sequences of concern" based entirely on their potential to be dangerous when incorporated into a synthetic construct, could theoretically facilitate the design of a synthetic pathogen by a "bad actor." Thus, the committee agreed with the NSABB that "The USG should include advances in synthetic biology and advances in our understanding of virulence/ pathogenicity in "tech-watch" or "science-watch" endeavors."

b. ***Danger of misimplementation:***

The committee stresses that a system for classification of Select Agents is based on classification and should not be confused with prediction. It would be a mistake to assume that "sequences of concern" or "parts" of Select Agents would function in a genome backbone (or context) independent manner. This information is partial and cannot be appropriately applied to designate a "sequence of concern" or individual "part" as a Select Agent. If the classification system were incorrectly implemented it would be counterproductive for security, and could be crippling to public health research. For example, many viruses encode a suite of "interferon antagonist genes" to target multiple steps in pathogen sensing, interferon signaling and innate/adaptive immunity. These genes are not themselves infectious or dangerous, and it isn't possible to forecast if these virulence genes/sequence motifs would enhance disease if introduced into a different genome back-

---

the properties to be covered under the Select Agents Regulations. How would this black box be used? Individuals would not know whether they possess or are transferring a Select Agent until the black box has been run on their genome sequence. Would we require that the black box be run on all new or modified genome sequences? This would surely be impractical; the nature of modern biology routinely involves innumerable modified or synthetic DNA constructs. But if we don't run the black box on *all* new or modified genome sequences, then we would need to define, with the clarity of a criminal statute, exactly who is required to run the black box and when. How big of a genetic modification requires a new black box certification? This would be a grey zone even worse than the problems associated with a simple Select Agents list. For new isolates of organisms from the wild, the genome sequence is not immediately known. Would individuals be required to obtain the genome sequence before commencing work on any new isolate?

bone. Designating such sequences as Select Agents would have little if any security benefit, but could have significant negative consequences by imposing undue burden on important vaccine research, e.g., influenza vaccines based on ns1 mutations/truncations and cell-surface glycoproteins, which are components of protective immunity and key for tissue tropism.

The yellow flag system would also be damaging to public health and security if misapplied. For instance, requiring registration or restricting access to flagged genomic fragments would become a significant barrier to scientific progress, including biodefense research. The yellow flag system would only be effective if maintained as a broad and flexible system for guidance and monitoring. The yellow flag system would provide valuable information to researchers, synthesis companies, and DNA hobbyists. Likewise, the yellow flag system could function as a nonintrusive information resource for law enforcement and the intelligence community.

c. ***Opportunity—Cost:*** A gene-sequence based oversight system, aimed at *prevention*, may not be the best use of resources. Such a system would be far from fail proof since a determined "bad actor" could produce synthetic DNA without the aid of synthesis companies; this would certainly include those operating outside of the United States.[7] Moreover, such an oversight system would be moderately expensive to implement in terms of both money and the time required of the highly skilled staff and expert advisors involved. The Commission on the Prevention of Weapons of Mass Destruction Proliferation and Terrorism recently concluded that "[d]eterrence of bioterrorism rests upon the ability of the nation to mitigate the effects of an attack," and that "the United States is woefully behind in its capability to rapidly produce vaccines and therapeutics, essential steps for adequately responding

[7]We must recognize that from the standpoint of impeding bioterror scenarios, there will be myriad ways to get around any screening procedure used by DNA synthesis companies, ranging from splitting an order into apparently innocuous pieces across multiple companies, to using offshore companies that do not adhere to U.S. regulations, to simply not using a DNA synthesis company at all. The technology of DNA synthesis is rapidly being commoditized, and DNA oligonucleotide synthesis machines can already be purchased cheaply from Ebay. An ebay.com search on "oligo synthesizer" on 10 October 2009 found a used Applied Biosystems 394 DNA/RNA oligo synthesizer on sale for $8,900 (plus $106.16 shipping within 3-8 business days to a committee member's home in Northern Virginia). With difficulty, genes and even whole genomes can be assembled from individual short oligonucleotides. In much the same vein, a determined bioterrorist can obtain isolates of a select agent from the wild. The SAR can only raise the difficulty bar for acquiring cultures of proven highly virulent agents, and provide law enforcement with tools to prosecute for possession of variants of such agents; because natural biological organisms are widely available, readily engineered, and increasingly easy to create, it is unrealistic to try to design the SAR to preclude acquisition completely.

to a biological threat, whether natural or man-made." Therefore, it is worth considering that, even in the context of national security, resources may be better used toward understanding infectious disease and developing response capabilities.

Appendix M

Executive Order: Optimizing the Security of Biological Select Agents and Toxins in the United States

For Immediate Release July 2, 2010

EXECUTIVE ORDER

OPTIMIZING THE SECURITY OF BIOLOGICAL SELECT AGENTS AND TOXINS IN THE UNITED STATES

By the authority vested in me as President by the Constitution and the laws of the United States of America, it is hereby ordered as follows:

Section 1. Policy. It is the policy of the United States that:

(a) A robust and productive scientific enterprise that utilizes biological select agents and toxins (BSAT) is essential to national security;

(b) BSAT shall be secured in a manner appropriate to their risk of misuse, theft, loss, and accidental release; and

(c) Security measures shall be taken in a coordinated manner that balances their efficacy with the need to minimize the adverse impact on the legitimate use of BSAT.

Section 2. Definitions. (a) "Select Agent Program" (SAP) means the regulatory oversight and administrative activities conducted by the Secretaries of Health and Human Services and Agriculture and the Attorney General to implement the Public Health Security and Bioterrorism Preparedness and Response Act of 2002 and the Agricultural Bioterrorism Protection Act of 2002.

(b) "Select Agent Regulations" (SAR) means the Federal regulations found in Part 73 of Title 42 of the Code of Federal Regulations, Part 331 of Title 7 of the Code of Federal Regulations, and Part 121 of Title 9 of the Code of Federal Regulations.

(c) "Biological Select Agents and Toxins" means biological agents and toxins with the potential to pose a severe threat to public health and safety, animal and plant health, or animal and plant products and whose possession, use, and transfer are regulated by the Department of Health and Human Services and the Department of Agriculture under the SAR.

Section 3. Findings. (a) The use of BSAT presents the risk that BSAT might be lost, stolen, or diverted for malicious purpose. The SAP exists to provide effective regulatory oversight of the possession, use, and transfer of BSAT that reduces the risk of their misuse or mishandling. The absence of clearly defined, risk-based security measures in the SAR/SAP has raised concern about the need for optimized security and for risk management.

(b) In addition, variations in, and limited coordination of, individual executive departments' and agencies' oversight, security practices, and inspections have raised concerns that the cost and complexity of compliance for those who are registered to work with BSAT could discourage research or other legitimate activities.

(c) Understanding that research and laboratory work on BSAT is essential to both public health and national security, it is in the interest of the United States to address these issues.

Section 4. Risk-based Tiering of the Select Agent List. To help ensure that BSAT are secured according to level of risk, the Secretaries of Health and Human Services and Agriculture shall, through their ongoing review of the biological Select Agents and Toxins List ("Select Agent List") contained in regulations, and no later than 18 months from the date of this order:

(a) designate a subset of the Select Agent List (Tier 1) that presents the greatest risk of deliberate misuse with most significant potential for mass casualties or devastating effects to the economy, critical infrastructure, or public confidence;

(b) explore options for graded protection of Tier 1 agents and toxins as described in subsection (a) of this section to permit tailored risk management practices based upon relevant contextual factors; and

(c) consider reducing the overall number of agents and toxins on the Select Agent List.

Section 5. Revision of Regulations, Rules, and Guidance to Accommodate a Tiered Select Agent List. Consistent with section 4 of this order, I request that:

(a) The Secretaries of Health and Human Services and Agriculture, no later than 15 months from the date of this order, propose amendments to their respective parts of the SAR that would establish security standards specific to Tier 1agents and toxins.

(b) The Secretaries of Health and Human Services and Agriculture each, no later than 27 months from the date of this order, promulgate final rules and guidance that clearly articulate security actions for registrants who possess, use, or transfer Tier 1 agents and toxins.

Section 6. Coordination of Federal Oversight for BSAT Security. To ensure that the policies and practices used to secure BSAT are harmonized and that the related oversight activities of the Federal Government are coordinated, the heads of executive departments and agencies identified in section 7(a)(ii) of this order shall:

(a) no later than 6 months from the date of this order, develop and implement a plan for the coordination of BSAT security oversight that:

(i) articulates a mechanism for coordinated and reciprocal inspection of and harmonized administrative practices for facilities registered with the SAP;

(ii) ensures consistent and timely identification and resolution of BSAT security and compliance issues;

(iii) facilitates information sharing among departments and agencies regarding ongoing oversight and inspection activities; and

(iv) provides for comprehensive and effective Federal oversight of BSAT security; and

(b) no later than 6 months from the issuance of final rules and guidance as described in section 5 of this order, and annually thereafter, review for inconsistent requirements and revise or rescind, as appropriate, any regulations, directives, guidance, or policies regarding BSAT security within their department or agency that exceed those in the updated SAR and guidance as described in section 5 of this order.

Section 7. Implementation. (a) Establishment, Operation, and Functions of the Federal Experts Security Advisory Panel.

(i) There is hereby established, within the Department of Health and Human Services for administrative purposes only, the Federal Experts Security Advisory Panel (Panel), which shall make technical and substantive recommendations on BSAT security concerning the SAP.

(ii) The Panel shall consist of representatives from the following, who may consult with additional experts from their department or agency as required:

— the Department of State;
— the Department of Defense;
— the Department of Justice;
— the Department of Agriculture (Co-Chair);
— the Department of Commerce;
— the Department of Health and Human Services(Co-Chair);
— the Department of Transportation;
— the Department of Labor;
— the Department of Energy;
— the Department of Veterans Affairs;
— the Department of Homeland Security;
— the Environmental Protection Agency;
— the Office of the Director of National Intelligence;
— the Office of Science and Technology Policy;
— the Joint Chiefs of Staff; and
— any other department or agency designated by the Co-Chairs.

(iii) To assist the Secretaries of Health and Human Services and Agriculture and the Attorney General in implementing the policies set forth in sections 1, 4,5, and 6 of this order, the Panel shall, no later than 4 months from the date of this order, provide consensus recommendations concerning the SAP on:

1. the designation of Tier 1 agents and toxins;
2. reduction in the number of agents on the Select Agent List;

3. the establishment of appropriate practices to ensure reliability of personnel with access to Tier 1 agents and toxins at registered facilities;

4. the establishment of appropriate practices for physical security and cyber security for facilities that possess Tier 1 agents. The Department of Homeland Security shall Chair a Working Group of the Panel that develops recommended laboratory critical infrastructure security standards in these areas; and

5. other emerging policy issues relevant to the security of BSAT.

Thereafter, the Panel shall continue to provide technical advice concerning the SAP on request.

(iv) If the Panel is unable to reach consensus on recommendations for an issue within its charge, the matter shall be resolved through the interagency policy committee process led by the National Security Staff.

(v) The Secretaries of Health and Human Services and Agriculture and the Attorney General shall report to the Assistant to the President for Homeland Security and Counterterrorism on the consideration and implementation of Panel recommendations concerning the SAP, including a rationale for failure to implement any recommendations.

(vi) The Panel shall be chartered for a period of4 years subject to renewal through the interagency policy committee process led by the National Security Staff.

(b) To further assist the Secretaries of Health and Human Services and Agriculture and the Attorney General in implementing the policy set forth in sections 1, 4, 5, and 6 of this order, the National Science Advisory Board for Biosecurity shall provide technical advice and serve as a conduit for public consultation, as needed, on topics of relevance to the SAP.

Section 8. Sharing of Select Agent Program Information. (a) Consistent with applicable laws and regulations, the Secretaries of Health and Human Services and Agriculture and the Attorney General shall, no later than 6 months from the date of this order, develop a process and the criteria for making SAP information available to executive departments and agencies when such information is necessary for furthering a public health, safety, security, law enforcement, or national security mission.

(b) SAP information shall continue to be safeguarded properly and handled securely to minimize the risk of disclosing sensitive, personal, and other information protected by the Privacy Act, 5 U.S.C. 552a.

Section 9. General Provisions. (a) The National Security Staff shall, on a biennial basis, review the implementation and effectiveness of this order and refer to the interagency policy committee process any issues that require further deliberation or adjudication.

(b) Nothing in this order shall be construed to impair or otherwise affect the authority granted by law to a department or agency, or the head thereof, or functions of the Director of the Office of Management and Budget relating to budgetary, administrative, or legislative proposals.

(c) This order shall be implemented consistent with applicable law and subject to the availability of appropriations.

(d) This order is not intended to, and does not, create any right or benefit, substantive or procedural, enforceable at law or in equity by any party against the United States, its departments, agencies, or entities, its officers, employees, or agents, or any other person.

BARACK OBAMA

THE WHITE HOUSE, July 2, 2010.